J. T. Saunders and S. M. Manton

A MANUAL OF PRACTICAL VERTEBRATE MORPHOLOGY

FOURTH EDITION

Revised by

S. M. Manton and M. E. Brown

CLARENDON PRESS · OXFORD

Oxford University Press, Ely House, London W.1

GLASGOW NEW YORK TORONTO MELBOURNE WELLINGTON
CAPE TOWN IBADAN NAIROBI DAR ES SALAAM LUSAKA ADDISABABA
DELHI BOMBAY CALCUTTA MADRAS KARACHI LAHORE DACCA
KUALA LUMPUR SINGAPORE HONG KONG TOKYO

FIRST EDITION 1931
SECOND EDITION 1949
THIRD EDITION 1959
FOURTH EDITION 1969
REPRINTED (with corrections) 1972

PRINTED BY OFFSET IN GREAT BRITAIN BY
WILLIAM CLOWES & SONS, LIMITED, LONDON, BECCLES AND COLCHESTER

*A Manual of
Practical Vertebrate Morphology*

Preface to the Fourth Edition

THE first edition of this Manual was designed in 1931 to fit the course for Part I of the Natural Sciences Tripos in Cambridge but the authors' hope that it would have a wider application proved to be justified. Six chapters describing dissections of muscles of selected vertebrates were added in 1949, bringing the second edition into line with the Cambridge course of that time. The third edition (of 1959) was enlarged by adding the general dissection of the rat, this animal having replaced the rabbit as the type generally used for mammalian dissection. The material added in 1949 and 1959 was placed at the end of the book and no major changes were made in the original chapters of the first edition.

In this fourth edition, the material used in the third edition has been rearranged so that descriptions of the muscles have been incorporated into the general dissections of the skate, pigeon, and rat. The whiting dissection has been changed into a chapter on Teleost fishes and includes detailed instructions for dissecting the trout and the roach in addition to the whiting. The chapters describing skeletons have been revised to include more about functions and expanded by adding sections on the crocodile and turtle. Brief accounts of the muscular systems of the head and trunk are now included.

Dr. S. M. Manton has drawn additional figures and Dr. M.E. Brown has been responsible for the detailed alterations and additions to the text. The classification used in the third edition of *Vertebrate palaeontology* by A. S. Romer (1966) has generally been followed.

Contents

Introduction

THE aim of this Manual is to provide the student with an accurate description of the anatomy of certain vertebrate types which are dissected in an elementary course of zoology. The description is supplemented by figures, all of which have been made by Dr. Manton. These figures have almost all been drawn from actual dissections or specimens. Both in the text and in the illustrations, only such details as can be seen by the ordinary methods of dissection are noted. The text follows the natural course of the dissection, the various parts being described in the order in which they should be seen by the student, if he is to make the fullest possible use of the material. The student is recommended to adopt this order, and to identify carefully the structures mentioned in one paragraph before passing on to the next.

In a manual of this nature space does not allow the insertion of details of comparative anatomy or much of the function of organs and structures. It has, however, been found necessary to provide a short general account of the vertebrate nervous system, in order that the arrangement of the nerves, particularly that of the cranial nerves, in accordance with their function, may be understood by the student. This arrangement is much more satisfactory than the older one of 10–12 cranial nerves (the nomenclature of which is retained for obvious reasons of convenience), but it suffers from the disadvantage that the student can in only a few cases discover by dissection the function of a nerve and the class to which it should be assigned.

The dissection of the types is followed by an account of the skeletons, which are here considered together for the sake of convenience, although, in practice, the student may study the skeleton at the same time as the dissection. In the description of the skeletons, and particularly of the skulls, other types in

addition to those dissected are described, for it is here both
possible and convenient to teach by comparative methods the
evolution of the skull and other structures. References to the
musculature have been included and the head muscles of
several animals are described with their skulls.

We have included in Chapters 14–16 a short account of the
main features of the osteology of the Mammalia, so that the
student may become generally acquainted with the variation
of structure in this group. Here, as elsewhere, the student in
reading the text is expected to have the actual specimen before
him.

Vertebrate dissection needs careful guidance if the student
is not to be discouraged by early mistakes. A mistake due to
lack of direction at the beginning too often ruins the single
prepared specimen, which is all that most laboratories can
provide for the student, and further progress then becomes
impossible. If the text is carefully read, and the figures con-
sulted, the student should have no difficulty in following a
course which will show all the principal features in the anatomy that can be shown by dissection (aided by a hand-lens),
and which will prevent him destroying structures and organs
before these have been properly examined. *The student is
recommended to colour the figures of the dissections with chalks
or crayons.* This will make the courses of the arteries, veins,
and nerves perfectly clear and will emphasize, in a way in
which a black-and-white drawing cannot, the distinction between the various structures shown. Care has been taken to
select for the paper on which this manual is printed one suit-
able for colouring in this way.

The muscular system is approached in this manual from
the point of view of function and it is hoped that the student
will try to investigate the actions of every muscle that he
observes during the course of each dissection. It must be
emphasized that the effects of contraction of a muscle cannot
be appreciated properly without some knowledge of the
skeleton; the student is therefore advised to refer constantly
to mounted or disarticulated skeletons while dissecting. If
this is done, certain skeletal features should be explicable in
terms of function and the form of the skeleton itself should
acquire a new interest. The actions that are attributed here
to muscles are those that can be observed during the dissection
of freshly killed animals. In living animals, the functions of
individual muscles may not be exactly those described here
because other muscles will be in different states of tension.

One of the difficulties encountered by students who are
interested in the muscular system is the problem of nomenclature. Too often, each group of vertebrates has its own series

of muscle names; different authors may use the same name for different muscles and different names for the same muscle. Ideally, the name of a muscle should give information about its position, its function, or its homologies, but homologies are often in doubt: one muscle may have several different functions according to the state of other muscles, and if names are given on the basis of position there may be a certain repetitive monotony about them which is discouraging to the beginner. In this account an attempt has been made to use a simple terminology by which muscles are grouped, generally according to function, and the groups are given distinctive names. Names in common use have been chosen where possible; the use of the same name in different animals is intended to suggest homologies, though it must be emphasized that the present work does not claim authoritatively to lay down homologies between the muscles of different animals. In general, the homologies for limb muscles suggested by Romer (1922, 1942, 1944) have been followed so that, for instance, the name *supracoracoideus* has replaced *pectoralis minor* (or *secundus*) in the pigeon. At the ends of the chapters the names used in this manual which differ from those of standard descriptions are equated with the latter. The student is advised not to attempt to learn and remember all the names but rather to use them as convenient labels.

The technique of dissecting muscles is in some ways simpler than that required for the dissection of nerves and blood-vessels. It consists of removing the connective tissue sheath from over a muscle and then separating that muscle along sheets of fascia (or connective tissue) from the surrounding structures. In a fresh animal the muscle should then be pulled and some idea of its action obtained before it is cut away from its origin and insertion. These are frequently on bones, but may be on fascia over other muscles, and they should be noted carefully as the muscle is removed. The dissections are easier to perform if each muscle is removed completely after it has been examined, but for purposes of display and drawing it is advisable to dissect out the more superficial muscles, to cut them away from either their origin or insertion, to reflect them outwards, and to draw the dissection before removing them completely and proceeding to the dissection of the deeper muscles. It is recommended that the muscles should be identified and investigated in the order described in the text. The student may find it helpful to colour the bones in the diagrams with chalks or crayons.

The dissection of the hypobranchial muscles of the skate has been introduced into the general dissection of the animal.

One side of the cod's head can be used for the dissection of the muscles and the other side for a dissection of the cranial nerves, as described for the whiting on pp. 48–54.

The investigation of the muscles of the frog, using stimulators, requires at least one practical period of two hours. If two periods are available, the student can learn to use a stimulator in the first period by examining the head muscles and may then dissect the viscera or vascular system. The second period can be used for the stimulation of the limb and girdle muscles of another frog.

If one pigeon only is available for the dissection of viscera as well as the muscles of the wing and leg, and a whole period is available for each of these three dissections, it is suggested that they should be done in the order: (1) wing, (2) viscera, (3) leg, and the bird should be kept fresh in a cold room between practicals. Alternatively, the wing and leg could both be dissected in one period, a second period being used for the dissection of the viscera.

The dissection of the fore-limb of the rat may precede a general dissection of the viscera and the dissection of the hind-limb can follow this. The animal should be kept fresh if possible.

If this manual makes it possible for the student to acquire more rapidly and easily than hitherto the necessary elementary knowledge of anatomy which must precede the study of physiology, then one of the chief aims of the authors will have been achieved.

ROMER, A. S. (1922) The locomotor apparatus of certain primitive and mammal-like reptiles. *Bull. Am. Mus. nat. Hist.* **16**, 517–606.
—— (1942) The development of Tetrapod limb musculature. The thigh of *Lacerta. J. Morph.* **71**, 251–98.
—— (1944) The development of Tetrapod limb musculature. The shoulder region of *Lacerta.* Ibid. **74**, 1–41.

NOTE

The words 'right (-hand)' and 'left (-hand)' always refer to the right or left side of the animal and not to the position of an organ as it may be seen in a ventral view.

1 *The lamprey*

(*Petromyzon fluviatilis*)

External features NOTE that the skin is everywhere smooth and devoid of scales. The division of the muscles into myotomes is usually apparent through the skin. The skin is slimy, due to the presence of numerous glands. In the middle line the skin is produced into two median dorsal fins, and one round the tail. These fins are stiffened by rods of cartilage: fin rays are absent. At the front end will be seen the circular mouth. Inside the rim of the mouth, which is used for attaching the animal to stones, will be observed a number of ectodermal horny teeth. The paired eyes have no eyelids. On the dorsal surface of the middle line is a single median opening which leads into the hypophysial cavity and the nasal sac. Behind this is a transparent area of skin above the pineal 'eye'. Behind the eyes on either side are the seven pairs of gill openings. The cloaca lies rather far back in the mid-ventral line. In most specimens there will be seen protruding from the mouth of the cloaca the slender urino-genital papilla. Note the absence of paired fins and the absence of a lateral line.

General directions for dissection Divide the animal *accurately* into two halves by a median longitudinal vertical cut with a razor or very sharp scalpel. The stiff gelatinous notochord can be seen extending from the head region throughout the body. Above it lies the nerve cord which dilates into the brain over the front end of the notochord and extends anterior to it. Over the nerve cord can be seen the cut myotomes or muscles. The body cavity containing the viscera occupies the major portion of the body. Anteriorly the body cavity is separated from the cardiac region by a delicate transverse membrane, the septum transversum. In front of this lies the pericardium and the branchial region of the body. Behind the body cavity the myotomes lie above the nerve cord and below the notochord.

**Anterior region
of the body**
(Fig. 1)

At the anterior end of the body note the buccal cavity of the mouth, well provided with horny ectodermal teeth. The entrance into the pharynx from the buccal cavity is dorsal. Immediately below this entrance will be seen a rasping tongue which is also provided with horny teeth. Projecting backwards from the tongue trace the very large muscles which serve to control its movement. Dorsal to the tongue trace the buccal cavity backwards to a point just below the brain. The cavity here narrows and divides into two. Ventrally there is an opening into the sub-oesophageal tube or pharynx. Note that the

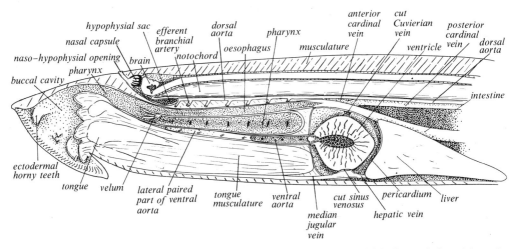

FIG. 1. Anterior part of the lamprey cut in the middle line and viewed from the median side.

opening into this tube is guarded by a fringe of delicate processes, the velum, which are attached to the upper side of the opening of the tube and hang downwards across it. The velum acts as a valve, allowing a pressure difference between the buccal cavity and the pharynx. Thus the lamprey can attach itself by suction pressure to a fish while still able to respire through the external gill openings. The pharynx extends backwards nearly as far as the heart. In its side walls will be seen seven openings, which lead into the branchial pouches. The openings of the branchial pouches to the exterior have already been seen in the examination of the external features. Immediately above the pharynx lies the oesophagus. The opening from the buccal cavity into the oesophagus lies immediately dorsal to the opening of the cavity into the pharynx; it is much narrower than this opening. It will usually be possible to trace the opening of the oesophagus into the buccal cavity by inserting a probe. Follow the oesophagus backwards.

In the region of the heart it passes dorsal to the pericardium and rather to the left-hand side. This passage of the oesophagus past the heart may be traced by inserting a probe. Behind the heart the alimentary canal dilates and will be seen lying in a groove on the left side of the liver. The alimentary canal now runs straight back to the anus. Note that the gut is about the same diameter throughout, that it is not twisted, and that it has no dorsal or ventral mesentery except at the posterior end where a very small dorsal mesentery can be distinguished. (See Fig. 3.) Note also that there is no distinction between the stomach and the intestine.

The liver is a large triangular body lying immediately behind the heart. Anteriorly it is closely applied to the septum transversum behind the pericardium, while dorsally it is joined directly to the wall of the alimentary canal as it enters the body cavity. There is neither gall-bladder nor bile-duct in the adult. The pancreas is very rudimentary, and is represented only by scattered packets of cells along the intestine. There is no spleen.

Open the intestine and find the typhlosole, which follows a loose spiral.

Vascular system
(Figs. 1, 2, 3)

The heart is an asymmetrical structure so that it will be necessary to examine both halves of the animal in order to determine all the relationships. The thick-walled ventricle lies to the right-hand side and fills up the major portion of the pericardial cavity. The thin-walled auricle lies to the left-hand side, and between it and the ventricle is the sinus venosus. From the ventricle the ventral aorta runs forward beneath the sub-oesophageal duct and above the musculature of the tongue. At its origin from the ventricle there is a pair of large pocket valves. In tracing the aorta forwards note the openings of the seven afferent branchial vessels.

Note that the ventral aorta is not an entirely median structure (see Fig. 1). It forks into two ventral aortae between the origin of the fourth and fifth afferent branchials. It may therefore be difficult to see the openings of the first four afferent branchial vessels in your specimen. If this is so, a little dissection will show these anterior afferent branchials as the two lateral aortae lie very close together.

The dorsal aorta is a median vessel lying immediately below the notochord extending the entire length of the body. The efferent branchial vessels, one from each gill pouch, are usually seen as cut stumps or pores.

Find the sinus venosus (Fig. 2). The sinus venosus is a very thin-walled structure lying between the auricle and the ventricle and extending from the dorsal to ventral side of the pericardium. Ventrally the sinus venosus receives the hepatic vein from the liver and the median jugular vein from the lower side of the tongue muscles. Dorsally the sinus venosus receives blood via the short single Cuvierian vein from the paired anterior and posterior cardinal veins. The Cuvierian vein runs vertically downwards on the right side of the intestine. The Cuvierian vein and the anterior and posterior cardinals are easily seen by inserting a probe into their cavities. Find also the sinu-auricular opening. Trace the posterior cardinals backwards and observe that they lie dorsal to the kidneys on either side of the dorsal aorta. Opposite the hinder border of the kidney they unite together and run backwards as a caudal vein (Fig. 3). Note that there is no renal portal system, the single, median caudal vein dividing to form the paired, posterior cardinals.

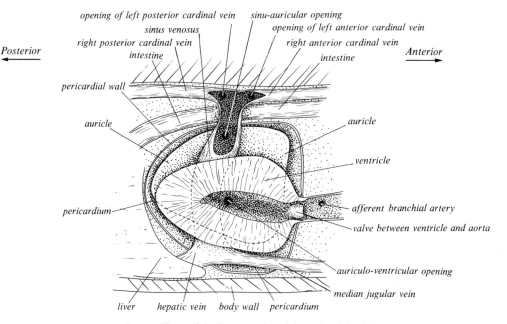

FIG. 2. Heart of the lamprey viewed from the right side.

Posterior region of the body
(Fig. 3)

The sexes are separate. The genital organs are elongated unpaired structures suspended from the dorsal wall of the body cavity by mesenteries. There are no genital ducts. The products of the gonads are shed directly into the body cavity and find their way to the exterior by the genital pores. These will be seen later after the examination of the kidneys.

The kidneys are suspended by mesenteries, attached to the body wall on either side of the gonad. They are elongated structures lying in the posterior part of the body cavity, and extend forwards about half or two-thirds its length. The ducts of the kidneys will be found on the lower border of each kidney. Trace one of these ureters forwards, when it will be found to extend beyond the anterior border of the kidney and run forwards ending near the pericardium. Trace the ureter backwards and note that just dorsal to the cloaca it unites with its fellow to form the urinogenital sinus. The urinogenital sinus is usually seen covered with a band of oblique muscle. In order to find the external opening of the ureter, slit up one ureter just anterior to the sinus and insert a bristle. Push the bristle backwards and see that it emerges at the tip of the urinogenital papilla.

Now find the genital pores. The openings of these from the body cavity into the urinogenital sinus are situated in the anterior lateral walls of the urinogenital sinus. In order to find these openings push a bristle backwards along the ventral inner surface of the body wall on one side or other of the rectum, between the rectum and the body wall. The bristle will find its way into the genital pores and emerge to the exterior by the urinogenital papilla.

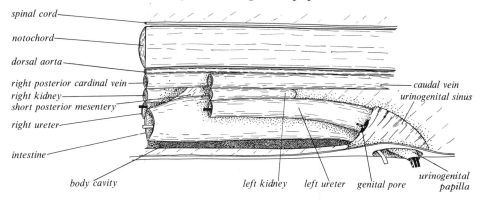

FIG. 3. Cloacal region of the lamprey dissected from the side. The left half of the body has been cut away leaving the lips of the cloaca and the urinogenital sinus intact. Black bristles are inserted through the ureters and are seen emerging from the urinogenital papilla. A black and white bristle passes from the body cavity through the genital pore to the urinogenital sinus and out through the urinogenital papilla. A white bristle passes through the intestine and projects from the cloaca. The caudal vein is seen bifurcating to form the posterior cardinal veins.

Slit open the rectum at its posterior end. Insert a bristle and find the opening of the anus into the cloaca.

The brain and spinal cord
(Fig. 1)

The brain in the lamprey is poorly developed and of small size. The most anterior part of the brain is the olfactory portion, which is in contact with the nasal sac. There are two small olfactory lobes, but the division between the olfactory lobes and the forebrain is not clearly visible. Behind the fore-brain there are two optic lobes, followed by a minute cerebellum. The cavity of the medulla oblongata can be clearly seen in the region posterior to the optic lobes together with the choroid plexus forming its dorsal roof. From the floor of the midbrain observe the minute pituitary body lying between the anterior end of the notochord and the posterior border of the nasal sac. The pineal body may be seen projecting from the dorsal side of the midbrain.

Trace the naso-hypophysial opening inwards (Fig. 1). Observe that it leads first of all into the median nasal sac lying in front of the brain; it then passes downwards and obliquely backwards under the nasal sac into a fair-sized cavity, lying immediately below the anterior end of the notochord and dorsal to the oesophagus. Note that this cavity, the hypo-physial cavity, is blind and has no communication with the oesophagus.

The hypophysial pit is formed from the dorsal wall of the stomodaeum and in the majority of the Craniata this hypo-physial pit comes into contact with the infundibulum from the floor of the midbrain to form what is known as the pituitary body. However, in the lamprey the growth of the upper lip in early development has carried the openings of the nasal and hypophysial pits to the dorsal side of the head. Here they unite but the hypophysial pit grows downwards and eventually reaches the ventral side of the brain, where it forms a pituitary body in the ordinary manner.

2 *The skate*

Raja naevus M. and H.

External features

General

OBSERVE that the animal is flattened dorsoventrally. Note that the spiracle is on the dorsal surface and that the gill slits are on the ventral surface. The dorsal surface is grey or brownish (protective coloration), the under surface is white. Compare the shape of the skate with the dogfish and ascertain how this flattening is produced.

On the dorsal surface note that the skin is rough and is covered with small placoid scales. There are also larger placoid scales with prominent spines distributed about the body and on the tail. (In the smooth skate, which is brownish in colour, there are no small placoid scales; only the larger ones are present.) The lateral expansions of the body are formed by the pectoral fins. Behind the pectoral fins are the pelvic fins (with claspers in the male). The tail starts between the pelvic fins, and narrows considerably backwards. Near the end will be seen two median dorsal fins, and immediately behind this is the caudal fin. The caudal fin is heterocercal, but it is poorly developed in comparison with the dogfish. At the anterior end of the body is the snout, which can be raised by means of two strong ligaments attached to muscles behind the head. (These ligaments will be seen later when dissecting the cranial nerves.) The eyes are large, and in the living animal are raised above the general surface of the body. Immediately behind the eye is the spiracle. Remove the mucus from the spiracle, and note the pseudobranch on its anterior border. The pseudobranch consists of folds of mucous membrane comparable to the gill filaments on the other branchial arches. At the anterior end of the animal, in the region of the snout, small pits can be observed, distributed as a rule in rows or lines between the scales. If the region near these pits is squeezed or pressed, mucus will be seen to be extruded from the pits. These pits are sensory structures, the ampullae of Lorenzini. They are supplied by branches of the lateral line nerves and can act as electric receptors. Details of the arrangement of these pits can be seen in Fig. 8.

The lateral line is a row of these sensory pits, each leading into a subdermal canal and supplied by a branch of the X (vagus) nerve. The external course of the lateral line cannot be traced except in the smooth skate, and only in this with difficulty. It lies just lateral to the mid-dorsal spines, and anteriorly branches to form supra- and infraorbital canals, etc.

Ventral surface

The mouth forms a transverse slit some way behind the snout, and, as in all Elasmobranchs, is not terminal. Separate the jaws and observe the teeth. These form a diamond pavement of modified scales. These teeth are not sharp, and are used for crushing the prey. (Compare with the dogfish.) Immediately in front of the mouth are seen the paired nasal openings, each of which is connected with the corner of the mouth by the oro-nasal groove. Behind the mouth are the five pairs of gill slits. Each slit has a fold of skin over it which acts as a flap valve. In respiration, the animal expands the pharynx and draws in water through the spiracle, the flap valves over the gill slits preventing the ingress of water here. The animal then contracts the pharynx and the water is expelled through the gill slits on the ventral surface.

Just behind the last gill slit, and towards the middle line, is a pair of transverse slits, leading into a shallow cavity in the skin. The function of these is unknown.

In the middle line between the pelvic fins is the cloaca. Pull aside the edges of the cloaca and observe the urinogenital papilla. (Unless this papilla is extruded, it will be difficult to find it in most specimens. It is as a rule much easier to observe in the male than in the female.) In the posterior lips of the cloaca will be seen the openings of the abdominal pores (see Fig. 4).

As on the dorsal surface, observe the lateral expanse of the pectoral fin. Immediately behind the pectoral fin is seen the pelvic fin, divided into two lobes. A clasper is present in the male. The clasper is only fully developed in large specimens, where the complicated apparatus for insemination can be seen.

General directions for dissection

Make a median incision through the abdominal wall from the anterior border of the pelvic girdle to the posterior border of the pectoral girdle. Then make transverse cuts, following the edges of the pectoral and pelvic girdles, to the side of the body cavity. Raise these two flaps, and look for the lateral abdominal vein, which lies underneath the peritoneum at the edge of the body cavity. After noting the position of the lateral abdominal vein, cut off the flaps at the sides of the body cavity. The contents of the body cavity will now be exposed. (In preserved specimens a portion of the abdominal body wall is usually removed in order to admit the preservative.)

Contents of the body cavity
(Fig. 4)

Arrange the contents of the body cavity as shown in Fig. 4, and identify in your specimen all the parts shown in the diagram. *Be careful not to tear or cut any mesenteries before these parts are identified.*

Alimentary canal

Underneath the liver observe the oesophagus, which here enlarges to form the U-shaped stomach. The cardiac portion of the stomach is large and forms one limb of the U. The pyloric portion is much smaller, and forms the other limb of the U. The stomach is separated from the intestine by the pyloric sphincter, the position of which is marked externally by a constriction followed by a thickening of the wall. The intestine is about as long as the cardiac portion of the stomach. Make a longitudinal slit in the middle of the intestine, and observe that it contains a spiral valve, as in the dogfish. Following the intestine is the rectum. The rectum has attached to it the rectal gland. Raise the rectal gland and observe the dorsal mesentery.

Liver, spleen, and pancreas

The liver is a trilobed structure, the median lobe being rather smaller than the other two. The gall-bladder lies exposed between the right and median lobes of the liver. With the forceps pick up the mesentery (gastro-hepatic omentum) joining the median lobe of the liver to the anterior portion of the intestine. The bile-duct lies uppermost in the ventral edge of this mesentery. It should be traced to its opening in the anterior end of the intestine. (Do not trace the bile-duct through the liver to the gall-bladder until you have observed the hepatic portal vein running in the mesentery immediately below (dorsal to) the bile-duct and entering the median lobe of the liver.) In this same mesentery lies the pancreas, a firm white gland consisting of two lobes—a large dorsal lobe, which is connected by a bridge of gland substance with the small ventral lobe, lying on the ventral side of the junction between the stomach and the intestine and closely connected to the gut by a mesentery.

The pancreatic duct enters the intestine from the posterior end of the ventral lobe of the pancreas. Cut open the intestine and find the internal opening of the pancreatic duct. (In some

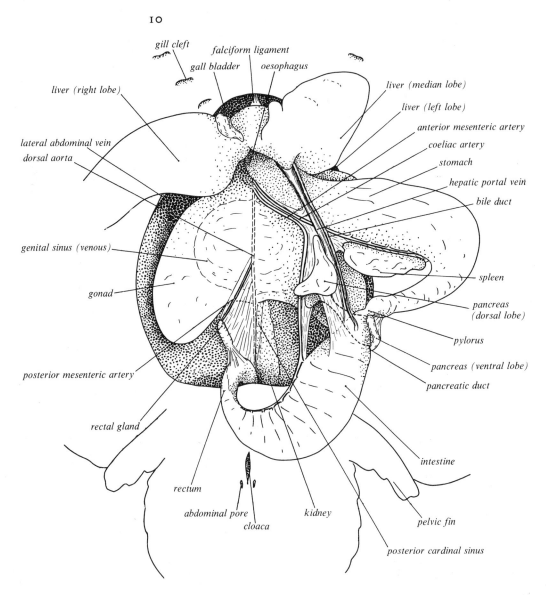

gill cleft

falciform ligament

gall bladder   oesophagus

liver (right lobe)

liver (median lobe)

liver (left lobe)

anterior mesenteric artery

coeliac artery

lateral abdominal vein

stomach

dorsal aorta

hepatic portal vein

bile duct

genital sinus (venous)

gonad

spleen

pancreas
(dorsal lobe)

pylorus

pancreas (ventral lobe)

posterior mesenteric artery

pancreatic duct

rectal gland

intestine

rectum

abdominal pore

cloaca

kidney

pelvic fin

posterior cardinal sinus

FIG. 4. View of contents of the body cavity of the skate after removal of the ventral body wall and before any dissection has been made. The liver is turned forwards.

specimens this opening is obscured by folds of the wall of the intestine.) The spleen is a dark red (brownish in preserved specimens) body, situated between the two limbs of the stomach and attached to the cardiac portion by a mesentery.

Trace the bile-duct through the liver to the gall-bladder.

In some specimens it will be possible to see hepatic ducts from the right and left hepatic lobes entering the bile-duct.

Arteries and veins

Turn the liver forward and pull the stomach well over to the animal's left. Close to the dorsal wall of the oesophagus will be seen running side by side the coeliac and anterior mesenteric arteries (Fig. 4). The anterior of the two is the coeliac artery. It has the following branches: a hepatic branch to the liver, an anterior gastric branch which divides into dorsal and ventral gastric arteries to the stomach wall, splenic branches to the spleen, a gastro-duodenal branch from which arise a posterior artery to the posterior part of the stomach, branches to the pancreas, and a duodenal branch to the pylorus and duodenum (anterior part of the intestine).

The anterior mesenteric artery gives off small branches to the pancreas and spleen and then goes to the intestine. Raise the rectal gland so as to stretch the dorsal mesentery, and observe the posterior mesenteric artery near the ventral border. The posterior mesenteric artery sends branches to the gonads and then passes to the rectal gland.

The hepatic portal vein drains blood from the walls of the stomach and intestine where the branches may be traced. As a rule these branches run alongside the arteries. The hepatic portal vein lies in the gastro-hepatic omentum dorsal to the bile duct. Trace it forwards to the middle lobe of the liver. Cut open the vein and trace its branches with a seeker to the right and left lobes of the liver.

Genital organs

Lying in the dorsal portion of the body cavity are the paired gonads, attached to the mid-dorsal line by mesenteries. The gonads in small specimens are indistinguishable as ovary or testis. Towards the middle line observe the large venous genital sinus. (Further examination and dissection of the urino-genital system will be carried out later.)

The hypobranchial muscles (Fig. 5)

The muscles situated below the pharynx of fishes are concerned with the movements of the branchial arches in feeding and respiration. The dorso-ventrally flattened form of the skate is associated with differences in the form of the branchial arches compared with those of the dogfish (notably the large median copula and the weak basihyal) and with differences in respiratory movements. The water inspired by the dogfish enters through the mouth and through the spiracle, but in the skate water enters only through the spiracle; in both fish water is expired through the gill clefts. This difference in the inspiratory current is marked by the presence in the skate of a spiracular pump, operated by movements of the hyomandibular cartilage and of the spiracular valve. The muscles of the skate are not as simple as those of the dogfish. They are derived from two sources: (1) the ventral ends of the posterior head myotomes, innervated by the hypoglossal nerve, (2) the lateral plate musculature of the gill segments, innervated by cranial nerves V, VII, IX, and X. In the dogfish the former are a series of paired longitudinally-arranged muscles arising from the pectoral girdle and inserted on to the ventral ends of the gill bars, while the latter form a constrictor sheath beneath the skin as well as the intrinsic muscles of the gill bars. In the skate all these muscles are modified; the relative size of some of them varies according to the species of skate.

The hypobranchial muscles and afferent branchial arteries

Remove the skin from the region between the mouth and pectoral girdle and as far laterally as the gill slits. Carefully dissect away the connective tissue on one side, exposing the *depressor rostri*, which arises near the mid-ventral line and, after passing round the angle of the jaw ventral to the *adductor mandibulae* muscle, forms a long tendon extending into the snout. Free the muscle below and pull on it and observe the change in shape of the snout. Reflect it forwards and dissect away connective tissue at the angle of the jaws to expose the small *depressor mandibularis* inserted on to the angle of the jaws. Remove this and below it find the *depressor hyomandibularis*, arising from connective tissue and passing laterally and dorsally to insert on the hyomandibular. This muscle forms part of the mechanism of the spiracular pump. Reflect it laterally and then dissect the other side of the fish to the same stage.

Clear the large median *coracomandibularis*, arising from the pectoral girdle and inserted on the lower jaw, pull on it and observe that the mouth opens. The mouth is closed by the *adductor mandibulae* which has already been noticed and which pulls the upper and lower jaws together. Cut the coracomandibularis away from the jaw, reflect it backwards, and remove the connective tissue beneath it. Note the median thyroid gland, the basihyal cartilage with a small *coracohyoideus* muscle attached to it on each side, and the anterior end of the ventral aorta dividing to form the anterior afferent branchial arteries. Remove the coracohyoideus muscles, and note that they arise from the surface of large *coracohyomandibularis* muscles which are inserted on the hyomandibular and form part of the spiracular pump mechanism. Cut these muscles close to their insertions on the hyomandibulars and pull them out from below (i.e. dorsal to) the afferent branchial arteries; then cut them away from their origin in very strong connective tissue in the mid-ventral line. Part of the ventral

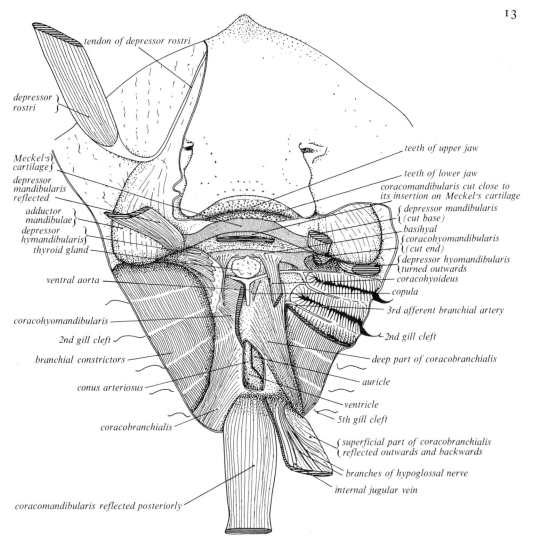

tendon of depressor rostri

depressor rostri

Meckel's cartilage
depressor mandibularis reflected
adductor mandibulae
depressor hymandibularis
thyroid gland

ventral aorta

coracohyomandibularis

2nd gill cleft

branchial constrictors

conus arteriosus

coracobranchialis

coracomandibularis reflected posteriorly

teeth of upper jaw

teeth of lower jaw
coracomandibularis cut close to its insertion on Meckel's cartilage
depressor mandibularis (cut base)
basihyal
coracohyomandibularis (cut end)
depressor hyomandibularis turned outwards
coracohyoideus
copula
3rd afferent branchial artery
2nd gill cleft
deep part of coracobranchialis
auricle
ventricle
5th gill cleft
superficial part of coracobranchialis reflected outwards and backwards
branches of hypoglossal nerve
internal jugular vein

FIG. 5. Partial dissection of the hypobranchial muscles of the skate (*Raja naevus* M. and H.). The skin has been removed from the branchial region, and the connective tissue between some muscles has been dissected away. Cut surfaces are shown by horizontal hatching and exposed cartilage by mechanical tint. In the middle line the coracomandibularis has been severed from Meckel's cartilage and reflected posteriorly. On the animal's right the depressor rostri and the depressor mandibularis are reflected forwards. On the animal's left the depressor mandibularis and the depressor hyomandibularis are cut short and reflected forwards and outwards, the coracohyomandibularis is cut and pulled out leaving the basihyal cartilage and afferent branchial arteries undisturbed, and the superficial part of the coracobranchialis is reflected outwards and backwards. A window has been cut through the pericardial floor exposing part of the heart. The division of the ventral aorta into the pair of anterior afferent branchial arteries is shown; the left anterior branchial artery is dissected out to show its division to form the first and second afferent branchial arteries passing to the gill arches, and the distal part of the third afferent branchial artery is similarly exposed.

aorta and the *coracobranchialis* muscles are now exposed; the latter are seen arising laterally from the pectoral girdle and passing dorsally towards the mid-line. The superficial part is inserted on the coracohyomandibularis and the deep part on the copula and there are small slips to the hypobranchial cartilages. This mass of muscles lowers the floor of the pharynx during inspiration. Remove the superficial part and note on its dorsal surface the branches of the hypoglossal nerve; cut backwards along the ventral aorta to the pericardium separating the deep coracobranchiales along the mid-line. Now carefully dissect these away from the copula, beginning at the anterior end and exposing the posterior afferent branchial arteries which arise as single vessels on each side from the aorta and each divide into three branches which pass through the muscle mass to the gills.

The vascular system

The afferent branchial system
(Fig. 5)

Now determine the blood-supply from the aorta to the gills. The first two (anterior) gill arches are supplied by the anterior afferent branchial artery which divides into two, and the posterior afferent branchial artery supplies branches to the three posterior gill arches. Trace the afferent branchial arteries along the gill arches, which should be separated from each other by slits made with scissors into the constrictor muscles. Observe the arrangement of the gill lamellae.

It will now be necessary to make out the relations of various organs and parts lying in the region of the falciform ligament which suspends the liver in the middle line from the anterior wall of the body cavity (Fig. 6). Carefully remove the middle portion of the pectoral girdle without injuring the liver. In the female, follow the oviducts from the posterior part of the body cavity forwards, and find the internal openings lying on either side of the falciform ligament and ventral to the oesophagus. If your specimen is a male, examine first a female specimen, then in your specimen note that rudimentary anterior portions of the oviducts can be seen.

Blood sinuses
(Fig. 6)

The hepatic sinus is a transverse vessel, lying against the anterior edge of the liver, ventral to the oesophagus close to the internal opening of the oviducts. Slit open the hepatic sinus and explore it with a seeker. It extends antero-laterally on either side, and opens into the Cuvierian veins (not directly into the sinus venosus as in dogfish). The cut end of the lateral abdominal vein can be found within the peritoneum of the anterior part of the body cavity lying against the pectoral girdle. Insert a seeker forwards into the lateral vein and inwards through the Cuvierian vein into the sinus venosus. Leave the seeker inserted and, with a scalpel, slice along it, cutting away the body wall and cartilages so as to expose the lateral abdominal and Cuvierian veins. Follow the lateral vein backwards, find two brachial veins joining it immediately behind the cut cartilages of the girdle and the fin respectively.

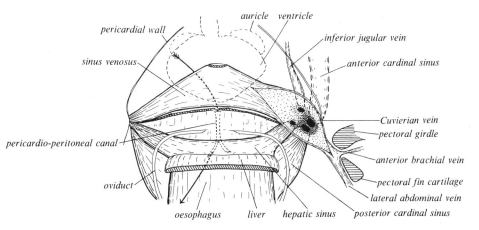

FIG. 6. Diagram showing the veins round the heart and the pericardio-peritoneal canal in the skate. The positions of the auricle and ventricle are indicated by the dotted lines. The left Cuvierian vein is cut open to expose the openings of the veins leading into it. The arrow passes through the pericardio-peritoneal canal.

Posteriorly the lateral vein arises from capillaries over the rectum and cloaca, and receives the iliac vein posterior to the pelvic fin cartilage. Examine the inner walls of the Cuvierian vein. Anteriorly it receives the inferior jugular vein, which runs in the dorso-lateral walls of the pericardial cavity and drains the mouth, pharyngeal, and pericardial regions. Probe into the Cuvierian vein in a dorsal direction. Two openings will be found: (1) an outer one that leads into the anterior cardinal sinus, which may be located by feeling for the end of the probe on the dorsal side; (2) an inner opening leads to the posterior cardinal sinus. Ventral to this opening lies the smaller opening of the hepatic sinus. Pass a seeker from the Cuvierian vein into the hepatic sinus. Probe into the posterior cardinal sinus from the Cuvierian vein, and cut open the sinus in the body cavity. The two posterior cardinal sinuses lie on either side and dorsal to the oesophagus. Trace them backwards to their origin on the inner side of each kidney. With a seeker find the connection between the posterior cardinal sinus on either side, and from this point pass the seeker laterally into the genital sinus. The posterior cardinal sinus receives blood from the kidneys, which are supplied by the renal portal veins formed by the bifurcation of the caudal vein.

Visceral branches
of the vagus nerve

The visceral branch of the vagus nerve lies along the sides of the oesophagus in the anterior part of the body cavity. Trace the branches from this nerve backwards to the stomach.

The heart
(Fig. 6)

Cut through the base of the aorta where it joins the conus arteriosus and turn the heart backwards. Insert a seeker in the middle line between the dorsal pericardial wall and the sinus venosus into the pericardio-peritoneal canal, and follow this canal backwards. It is about 1 inch long, lying on the ventral wall of the oesophagus. It is quite transparent, and the probe will be seen passing along it and finally coming out into the body cavity. Actually the opening of the canal into the body is through a number of minute pores at the end of the canal, and the probe in passing out breaks through the porous ending. Cut through the junction of the Cuvierian veins with the sinus venosus and remove the heart from the pericardium into a small dish. Slit open the heart by a median longitudinal ventral incision and examine the valves.

Embryologically the vascular system is formed from a pair of dorsal and a pair of ventral vessels. The paired dorsal vessels partially fuse together to form the dorsal aorta, while ventral aorta, heart, hepatic veins, and the hepatic portal vein are formed by the partial fusion of the paired ventral vessels. The heart arises as an S-shaped twist in the ventral vessel and becomes divided into four chambers. These chambers are originally in a straight line but become twisted, thus packing the chambers closer together. The heart lies in a special portion of the body cavity, the pericardial cavity. The posterior dorsal chamber of the heart is the sinus venosus forming a triangular sack extending across the posterior part of the pericardial cavity. Laterally it receives blood from the Cuvierian veins. Anteriorly it opens into the auricle through a median opening which is guarded by the sinu-auricular valves. The auricle lies in the dorsal anterior part of the pericardial cavity and opens posteriorly into the ventricle. The entrance into the ventricle is guarded by the auriculo-ventricular valves. The ventricle is thick-walled and occupies the major part of the pericardial cavity. The conus arteriosus leaves the anterior end of the ventricle and penetrates through the pericardium anteriorly in the middle line. The conus arteriosus bears three longitudinal rows of pocket valves, with five valves in each row.

The dorsal arterial system (Fig. 7)

With a strong pair of scissors cut through the angles of the jaws and of the gill arches on either side. Remove the floor of the pharynx and pericardium. Remove the mucous membrane from the roof of the mouth as far as the liver. Carefully

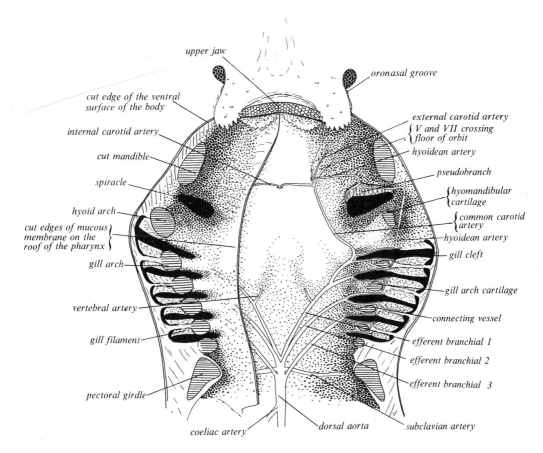

FIG. 7. Dissection of the efferent branchial system of the skate. The lower jaw and ventral parts of the gill arches and the floor of the pharynx have been removed, and the mucous membrane from the roof of the mouth and pharyngeal region has been stripped away on the left side exposing the blood-vessels; part of the left hyomandibular cartilage has been cut out.

remove the liver, leaving the rest of the viscera intact. The dorsal aorta will be seen behind the pectoral girdle in the middle line. It is formed by the union of three main pairs of efferent branchial vessels. Trace these vessels forwards and outwards to the branchial arches, removing the pharyngobranchial cartilages where necessary. The first is formed by the union of two efferent branchial arteries from the first two gill clefts. Find each of the four efferent arteries arising by the union of branches draining each side of one gill cleft. The anterior hemibranch of the fifth cleft drains into the vessel from the posterior hemibranch of the fourth cleft. From the first main efferent artery the vertebral artery arises and, passing inwards and forwards, it disappears through a foramen into the vertebral column. At the point of departure from the gill cleft of the most anterior efferent branchial there arises the common carotid artery. It passes forwards across the base of the hyomandibular cartilage along the ventral outer border of the cranium. Just anterior to the spiracle the common carotid artery divides into internal and 'external' carotids.[1] The internal carotid artery turns at right angles towards the middle line and, with its fellow from the opposite side, enters the cranium through a median foramen. The 'external' carotid artery passes forwards across the floor of the orbit. Just in front of the origin of the internal carotid artery the hyoidean artery can be seen crossing over the external carotid to enter the cranium by a lateral foramen. Follow this vessel outwards to the pseudobranch and round the outer border of the spiracle. From here it runs backwards to arise from the vessel draining the anterior hemibranch of the first cleft. To expose the course of this vessel between the first cleft and the spiracle the outer end of the hyomandibular cartilage must be removed.

Trace the dorsal aorta backwards. Just before the last efferent branchial enters the dorsal aorta the subclavian arteries arise on either side. In its course towards the tail the dorsal aorta gives off in succession the coeliac, the anterior mesenteric, the genital, and the posterior mesenteric and renal arteries. The course of some of these arteries has already been described (see p. 11).

The urinogenital system

Remove the body wall in the posterior part of the abdomen and cut away the pelvic girdle. Remove one pelvic fin without damaging the cloacal region.

Insert a probe anteriorly through the external opening of the abdominal pore and see it emerge in the posterior part of the body cavity.

The genital system is not fully developed except in large specimens.

---

[1] The 'external carotid' artery of Elasmobranchs is not homologous with the external carotid of higher forms. In the latter the external carotids represent the anterior ventral roots of the ventral aorta. These are absent in Elasmobranchs since the ventral aorta ends blindly after giving off the first afferent branchial arteries. The so-called 'external carotids' are vessels formed from the dorsal arterial system.

Male

The testes have already been seen in the examination of the contents of the body cavity (see p. 11).

Examine the mesentery connecting the testis with the anterior (sexual) portion of the kidney and find numerous fine vasa efferentia which connect the ducts in the testis with those in the kidney.

Turn the testis over to one side and remove the peritoneum from the surface of the kidney. Observe at the anterior end of the kidney, which lies on either side of the oesophagus, the large coiled Wolffian duct forming the epididymis. Follow the duct backwards and note that it is dilated into a sperm sac, just before it enters the cloaca. Cut open the posterior end of Wolffian duct and insert a probe; observe that this finds its way into the cloaca by the urino-genital papilla. Dorsal to the Wolffian duct lies the kidney. The anterior (sexual) portion is elongated and smaller in bulk than the posterior (excretory) portion of the kidney.

Note that the posterior portion of the kidney lies in a backwardly projecting pocket of the body cavity on either side of the cloaca. On the inner side of the posterior (excretory) portion of the kidney lies the ureter dorsal to the Wolffian duct. Pull aside the Wolffian duct and trace the ureter forwards and observe that it is formed by branches draining the posterior (excretory) portion of the kidney. Trace it to its connection with the cloaca. Note that the Wolffian ducts and the ureters on either side unite to form a small urinogenital sinus which opens into the cloaca by the urinogenital papilla.

Examine the inner side of the posterior (excretory) portion of the kidney and find small yellowish inter-renal bodies. These inter-renal bodies are composed of similar tissue to the cortex of the mammalian adrenal body. (Supra-renal bodies are present lying on the course of the sympathetic nerve-chains. In higher forms the homologues of these supra-renals form the medulla of the adrenal body.)

In the immature males, as often provided for dissection, the anterior sexual portion of the kidney is very narrow, and the Wolffian duct is small and straight throughout.

Female

The ovaries are paired, not single as in dogfish. They lie in similar positions to the testes and are of the same shape. The internal opening of the oviducts has already been found (see Fig. 6). Trace the oviducts backwards from the internal opening. The more anterior portion is thin-walled and slender. Near the head of the kidney the oviduct widens. The anterior end of this widening forms a solid body called the oviducal or shell gland. This gland secretes the horny covering in which the eggs are laid. It is only developed in fully matured specimens.

Posteriorly the two oviducts open into the anterior end of the cloaca by separate openings. Dissect off the peritoneum from the ventral surface of the kidney. Note that the kidney

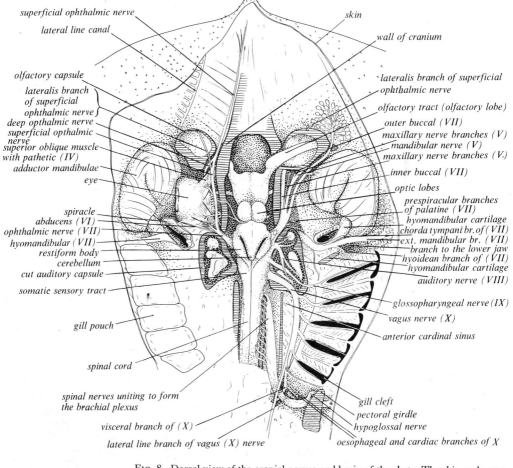

superficial ophthalmic nerve

lateral line canal

skin

wall of cranium

olfactory capsule

lateralis branch of superficial ophthalmic nerve

deep opthalmic nerve
superficial opthalmic nerve
superior oblique muscle with pathetic (IV)
adductor mandibulae
eye

lateralis branch of superficial ophthalmic nerve

olfactory tract (olfactory lobe)

outer buccal (VII)
maxillary nerve branches (V)
mandibular nerve (V)
maxillary nerve branches (V.)

inner buccal (VII)

optic lobes

spiracle
abducens (VI)
ophthalmic nerve (VII)
hyomandibular (VII)
restiform body
cerebellum
cut auditory capsule

somatic sensory tract

gill pouch

spinal cord

spinal nerves uniting to form the brachial plexus

visceral branch of (X)

lateral line branch of vagus (X) nerve

prespiracular branches of palatine (VII)
hyomandibular cartilage
chorda tympani br. of (VII)
ext. mandibular br. (VII)
branch to the lower jaw
hyoidean branch of (VII)
hyomandibular cartilage
auditory nerve (VIII)

glossopharyngeal nerve (IX)

vagus nerve (X)

anterior cardinal sinus

gill cleft
pectoral girdle
hypoglossal nerve

oesophageal and cardiac branches of X

FIG. 8. Dorsal view of the cranial nerves and brain of the skate. The skin and some connective tissue have been removed from the animal's left side, and on the right the eye and its muscles have been removed and the dissection carried further. The cut surfaces of cartilage are cross-hatched. Not all structures are labelled. The forebrain lies in front of the optic lobes, and below it on either side the optic nerves can be seen. Behind the superior oblique muscle lie three other eye muscles, the internal, superior, and external rectus muscles. The III nerve is seen emerging below the pathetic (IV) from the brain, and passes through the cranium where it divides. The two anterior branches supply the internal and superior rectus muscles, and the third branch turns downwards between the superior and external rectus to supply the inferior rectus and inferior oblique muscles. The brachial nerve is seen passing through the pectoral girdle cartilage on the right side.

only extends forwards about half the length of the body cavity. Note that the anterior (sexual) portion of the kidney in the male is here absent and the entire kidney is excretory.

As in the male find the ureter and trace it backwards to the cloaca. The ureter unites with its fellow to form a small urinary sinus situated on the dorsal surface of the anterior end of the cloaca. The urinary sinus opens into the cloaca between the openings of the oviducts. The inter- and supra-renals are the same as in the male.

The nervous system and sense organs
(Fig. 8)
General directions and superficial nerves

Remove the skin between the eyes and over the median region of the snout and over the branchial region. Below the skin notice the canals joining the lateral line sense pits, the openings of which to the surface have already been seen in the examination of the external features. Find the *levator rostri* muscle arising from fascia above the gill region and inserting by a long tendon at the side of the snout. Carefully cut away the cartilaginous roof of the cranium so as to expose the brain. Care must be taken at all stages of the dissection not to damage the brain. Find the superficial ophthalmic nerves (V and VII) emerging through the anterior cartilaginous wall of the orbit and running forward over the nasal capsule to the anterior part of the snout. Pare away the cartilage until the superficial ophthalmic nerve is now exposed between the cranium and the olfactory capsule. Over the olfactory capsule the superficial ophthalmic joins the deep ophthalmic nerve. Just posterior to the junction of the deep and superficial ophthalmics the superficial ophthalmic gives off a branch which runs outwards across the olfactory capsule. Trace the superficial and deep ophthalmics forward to the tip of the snout, and note that these nerves end in very fine processes just beneath the skin. Next trace the branch of the superficial ophthalmic running across the olfactory capsule (see Fig. 8). It will be seen to supply a group of lateral line sense organs just in front of the *adductor mandibulae* muscle.

You have now seen that the superficial ophthalmic is not a simple but a mixed nerve. A portion of it runs forward on to the snout to supply sense organs in the skin; this is the somatic sensory portion, while another portion runs across the olfactory capsule to the special lateral line sense organs and is known as a lateralis nerve. It will be necessary for you to trace separately all the branches of the cranial nerves to their destinations in order to discover, so far as is possible, what their functions are. You will find that the V, VII, IX, and X nerves are all 'mixed' nerves.

*The eye, its muscles,
and their nerve
supply*

The eyeball is moved by six muscles arranged in two groups attached to the anterior and posterior inner angles of the orbit. The anterior group consists of two oblique muscles. The superior oblique muscle is attached to the dorsal anterior part of the eyeball and the inferior oblique muscle is attached to the ventral anterior part. The posterior group of muscles consists of four recti muscles. The superior rectus muscle is attached to the posterior part of the eyeball, the internal rectus to the dorsal inner part of the eyeball, the external rectus to the posterior side of the eyeball, and the inferior rectus to the median lower edge of the eyeball.

Before proceeding any further find the nerves supplying the eye and its muscles. These nerves are as follows:

    II. the optic nerve supplying the retina of the eye,

    III. the oculo-motor supplying the superior, inferior, and internal recti and inferior oblique muscles,

    IV. the pathetic supplying the superior oblique muscle,

    VI. the abducens supplying the external rectus muscle.

Consult Fig. 8 and look for the nerves within the cranium. The IV nerve (pathetic) arises from the dorsal side of the brain between the cerebellum and optic lobes and proceeds forwards to its foramen in the wall of the cranium. Immediately below the foramen, through which the IV nerve passes, find the large II (optic) nerve passing directly outwards to the orbit. Behind the optic nerve can be seen the III (oculo-motor) nerve arising from the ventral side of the brain and passing to its foramen in the wall of the cranium. Remove the eyelids and pare away the wall of the cranium so as to expose the superficial ophthalmic nerve lying dorsal to the eye muscles, taking care not to destroy the IV nerve as it passes through the wall of the cranium.

Gently pull the eye outwards to stretch its muscles. Observe the IV nerve passing forwards and outwards to the superior oblique muscle. Then pare away the wall of the cranium behind the foramen of the IV nerve until the root of the ophthalmic nerve is exposed passing through the cranium. In the posterior part of the orbit look for the III nerve as it emerges through the cranium behind the IV nerve. Two branches will be seen to leave the III nerve as it enters the orbit going to the internal and superior recti muscles. The III nerve then curls downwards between the superior and external recti muscles and supplies the inferior rectus and inferior oblique muscles. Turn the eye upwards and examine its lower surface. A long branch of the III nerve will be seen emerging between the recti muscles and passing forwards to the inferior oblique muscle.

Examine the lower edge of the posterior rectus muscle and find the small VI (abducens) nerve supplying it.

Find the large II (optic) nerve emerging from its foramen through the wall of the cranium and entering the eyeball.

The nerves lying across the floor of the orbit

Remove the eye by cutting through the recti and oblique muscles and the optic nerve close to the wall of the cranium (for dissection of the eye see p. 28). On the floor of the orbit will be seen three large nerves emerging together from the cranium at the posterior corner of the orbit. The outermost of the three nerves is the outer buccal and the inner one is the inner buccal (VII). Between them lie the maxillary and mandibular nerves (V). The larger of these middle nerves close to its origin gives off a slender maxillary branch from the median side, and continues across the floor of the orbit as the mandibular nerve.

Trace the outer buccal nerve forwards and outwards to the anterior border of the adductor mandibulae muscle and then to the olfactory capsule. It passes through the lateral part of the capsule, and, after emerging from its foramen, it terminates in a group of lateral line ampullae. Note that each ampulla is separately supplied by a fine terminal branch of the nerve.

Trace the mandibular nerve (V) lying next to the outer buccal. It follows the course of the outer buccal as far as the adductor mandibulae muscle. On reaching the muscle it turns downwards and curls backwards along the cartilage of the upper jaw below the adductor muscle. Turn the animal over and dissect away the skin from the angle of the jaw. Find the nerve again and trace it backwards to the angle of the jaws where it curls round and supplies the lower jaw. This nerve is a visceral motor nerve and is the main supply of the adductor muscle. One of the branches to the adductor muscle can be seen leaving the mandibular nerve at the point where it first reaches the muscle.

On the inner border of the mandibular nerve (V) near the point where it reaches the adductor muscle will be noticed several small branch nerves. These nerves and the branch already noticed leaving the base of the mandibular nerve constitute the maxillary nerves (V), which pass anteriorly to supply the skin in the region of the oro-nasal grooves. These are somatic sensory nerves.

Trace the inner buccal forwards. This nerve on reaching the olfactory capsule passes ventrally downwards and may be found again on the ventral side of the body where it runs forward to supply the lateral line sense organs on the ventral side of the snout.

The VII and VIII nerves

If this has not already been done, carefully remove the skin round the spiracle. Look for the hyomandibular branch of the VII nerve which lies very close to the surface midway along the posterior border of the spiracle. Trace this nerve outwards. It runs past the posterior border of the aductor muscle as the external mandibular nerve (VII), and almost immediately after passing the adductor muscle it breaks up to supply a large group of lateral line ampullae. At the point where the external mandibular passes the muscle it gives off a nerve which runs round the outer edge of the adductor muscle, and may be traced forwards to the snout where it supplies the sense organs of the skin (somatic sensory). Now proceed to trace the remaining branches of the exposed portion of the VII nerve. The chorda tympani nerve branches from its anterior border just behind the spiracle. It follows the course of the external mandibular as far as the hyomandibular cartilage, where it turns ventrally and runs inwards and downwards between the adductor muscle and the cartilage, emerging from the ventral side immediately behind the angle of the lower jaw. It then runs inwards following the posterior border of Meckel's cartilage. The next branch is the hyoidean nerve, which arises from the posterior border of the VII nerve about $\frac{1}{8}$ inch distal to the origin of the chorda tympani. It runs downwards behind the hyomandibular cartilage, to supply the musculature of the hyoid arch, after which it continues inwards to supply the muscles of the floor of the pharynx. About $\frac{1}{4}$ inch beyond the origin of the hyoidean nerve a small branch leaves the VII nerve and passes downwards, following the posterior border of the adductor muscle to the ventral surface, where it crosses the adductor muscle immediately below the skin and runs to the anterior border of Meckel's cartilage.

Trace the VII nerve inwards to the point where it dips below the auditory capsule. After this it will be necessary to remove the skin, muscles, and cartilage from the posterior part of the cranium and the anterior part of the vertebral column. In order to do this, hold the fish in the left hand and place the middle finger on the point of junction of the skull and vertebral column. Depress the snout with the thumb so as to produce flexure of the brain in the region of the medulla. Now take a sharp scalpel and slice horizontally backwards through the muscles and cartilage so as to expose the medulla and first half-inch of the spinal cord. The roots of the VIII, IX, and X nerves will now be exposed.

The VIII (auditory) nerve enters the auditory capsule at its inner anterior corner. Note the three semicircular canals and the membranous labyrinth of the inner ear in slicing through the capsule. In the anterior part of the capsule the VIII nerve sends branches to the semicircular canals and ampullae. In the posterior part of the capsule two branches of the VIII nerve run backwards and outwards to the membranous labyrinth of the ear. Notice the IX nerve entering the auditory capsule a short distance behind the VIII nerve. It runs underneath the posterior branches of the VIII nerve and crosses the capsule to emerge from its outer posterior corner. Slice away the front part of the auditory capsule going down deeper so as to expose the root of the VII nerve from the brain. Close to the origin of the VII nerve from the cranium find the palatine nerve running forwards below the inner buccal nerve on the floor of the orbit. The prespiracular branches leave the base of the palatine nerve and pass outwards to the anterior wall of the spiracle. Follow the palatine nerve forwards above the mucous membrane of the roof of the mouth.

The IX and X nerves

The origin of the IX nerve has already been described. Pick up the IX nerve where it leaves the outer wall of the auditory capsule. Trace it outwards and note the two main branches running on either side of the first gill cleft.

The X (vagus) nerve arises immediately behind the IX. It arises from the brain by a number of roots, and runs outwards through the posterior wall of the auditory capsule. On leaving the auditory capsule it turns backwards and lies on the wall of the anterior cardinal sinus. Trace the branches of the vagus nerve to the gill arches. There are four of these branches and each divides to pass on either side of each gill cleft. Half way down the cardinal sinus the lateral line branch arises from the upper edge of the vagus. On its course backwards the lateral line nerve gives off branches to supply various sets of lateral line sense organs distributed about the surface of the body. The first of these may be seen just behind the last gill arch. At the posterior end of the cardinal sinus find two small branches leaving the vagus nerve and passing outwards and downwards in front of the pectoral girdle. Follow these branches downwards. The anterior branch passes to the ventral wall of the oesophagus. The posterior branch is the cardiac branch and supplies the heart. Behind the pectoral girdle the vagus nerve passes into the body cavity as the visceral branch and runs along the oesophagus to the stomach where its course has already been traced.

Behind the last cranial nerve or vagus will be seen the roots of the spinal nerves. These run outwards from the vertebral column and more or less unite to form a brachial plexus of nerves supplying the pectoral fins. Follow the brachial plexus backwards and outwards. It passes between the lateral line and visceral branches of the vagus to the cartilage of the pectoral girdle. Here it divides. The main branch continues through the cartilage as the brachial nerve which breaks up into branches to all parts of the pectoral fin after passing through the cartilage. The other branch forms the hypo-glossal nerve. It follows the anterior border of the pectoral girdle and passes downwards behind the last gill cleft. It then runs forward close to the skin along the side of the pericardium to supply muscles in the floor of the pharynx.

The brain
(Figs. 8, 9)

It will be convenient to examine the dorsal and ventral aspects of the brain and then its internal structure.

DORSAL ASPECT. At the anterior end observe the swollen forebrain partially divided into lateral portions by a shallow furrow. At the side of the forebrain the olfactory lobes pass outwards as a band of tissue lying across the olfactory capsule. The small and numerous olfactory nerves pass from the olfactory lobes to the olfactory capsules. Behind the forebrain note the two optic lobes partly covering the posterior part of the midbrain. Immediately in front of the optic lobes is the choroid plexus forming a thin roof to the third ventricle. Behind the optic lobes lies the cerebellum which covers the anterior part of the medulla oblongata posteriorly and partially overlaps the optic lobes anteriorly. At the sides the cerebellum is connected to the medulla by large restiform bodies (posterior cerebellar peduncles). Raise the posterior part of the cerebellum and notice the choroid plexus roofing the fourth ventricle.

Remove the brain from the cranium by cutting through the nerve-roots as close to the skull and as far from the brain as possible.

VENTRAL ASPECT (Fig. 9a). Note the floor of the forebrain. Immediately behind the forebrain will be seen the optic chiasma. Behind this is the infundibulum and pituitary body. On the floor of the midbrain behind the optic chiasma and on either side of the infundibulum are two larger lobes, the inferior lobes. Behind these lobes is a thin-walled hollow sac called the saccus vasculosus. It is bright red in the fresh specimens owing to the blood it contains. The under-surface of the midbrain behind the infundibulum (tuber cinereum) is formed by the crura cerebri. Behind the crura cerebri lies the medulla oblongata. It is broad anteriorly and narrows behind into the spinal cord. Ventrally it is marked by a median shallow furrow continuous with the ventral groove of the spinal cord.

Origins of the cranial nerves

The I (olfactory) nerves originate from the olfactory lobes which have already been described. The right optic nerve arises from the left side of the brain and the left optic nerve from the right side of the brain. Crossing over they form the optic chiasma below the floor of the midbrain. The III nerve arises from the floor of the midbrain through the inferior lobes.

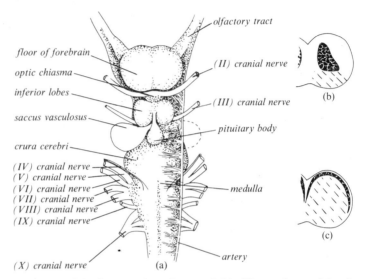

FIG. 9. (a) Brain of the skate seen from the ventral side. The arteries supplying the brain have been removed on the left side.
(b) Transverse section through the forebrain to show the ventricle and the thickness of the walls. The corpus striatum is cross-hatched.
(c) Transverse section through the forebrain of the whiting to show the ventricle and the thickness of the walls. The corpus striatum is cross-hatched.

The IV nerve arises from the roof of the midbrain between the optic lobes and the cerebellum. The V, VII, and VIII nerves have their roots inextricably mingled and their origin forms a conspicuous mass at the anterior lateral edge of the medulla. The VI nerve arises from the ventral surface of the medulla behind the pons. The IX nerve arises a short distance behind the origin of the V, VII, and VIII nerves. The X nerve arises by several roots from the hinder lateral part of the medulla.

Internal cavities of the brain

The internal cavities (or ventricles) of the brain can be seen in longitudinal sections. Make a vertical longitudinal para-sagittal cut through the brain slightly to the right or left of the middle line. One of the cavities of the forebrain will be exposed. Discover the connection of this cavity with the third ventricle. Note also the cavity of the fourth ventricle. The connection between the third and fourth ventricles will be seen in the next section. Make a vertical sagittal cut and then observe the third and fourth ventricles lying in the midbrain and hindbrain respectively. These two ventricles are connected by a narrow canal, the Sylvian aqueduct. Immediately anterior to the cavity of the fourth ventricle will be seen a narrow opening leading into the ventricle of the cerebellum. Note everywhere the thickness of the walls, particularly in the forebrain, where the walls of the floor and of the roof are practically equal in thickness (Fig. 9b). The dorsal roof of the third and fourth ventricle is devoid of nervous matter, and the covering is a thin vascular choroid plexus. These thin roofs are often damaged in the removal of the brain so that the ventricles appear to have no dorsal covering.

The eye

Remove the eye from the orbit. The external covering is the cartilaginous sclerotic. In the part of the eye that is exposed the sclerotic passes into the transparent cornea. The sclerotic is pierced by a small hole for the transmission of the optic nerve. Near this hole is a small prominence on the sclerotic for the articulation of the optic peduncle, a stalk of cartilage which articulates at one end with the eyeball and at the other end with the wall of the orbit just behind the optic foramen. Cut through the sclerotic and observe the next layer, the black choroid. At one point where the sclerotic joins the cornea the choroid becomes the iris and is coloured. The iris is pierced by a central aperture, the pupil. Immediately behind the iris and filling the aperture of the pupil is the lens. Lying on the choroid, from which it is easily detached, and forming the lining of the inside of the eyeball is a delicate membrane, the retina. The retina extends as far as the junctions of the choroid and the iris. Between the iris and the cornea is a space filled with a watery fluid called the aqueous humour. Behind the lens is the large posterior chamber, enclosed in the space between the retina and the inner surface of the lens and iris. The posterior chamber is filled with a gelatinous substance, the vitreous humour.

3 *The trout, the roach, and the whiting*

THIS chapter describes the brown trout (*Salmo trutta*) or rainbow trout (*Salmo gairdneri*) and has sections on the roach (*Rutilus rutilus*) and the whiting (*Gadus merlangus*).

The Teleost fishes have many features in common but also display much anatomical variety. Salmonids are considered to be primitive (super-order Protacanthopterygii) and fresh trout are convenient for dissection, especially of the viscera and blood system. The nervous system is best studied in preserved specimens and is described in detail here for the whiting, which belongs to the super-order Paracanthopterygii. This fish also illustrates some of the features of advanced Teleosts. The roach is an example of the successful fresh-water super-order Ostariophysi, of which Weber's ossicles are a diagnostic character.

In this chapter, for each stage of observation and dissection the trout is described first, followed by the roach and then the whiting. The central nervous system is described in detail for the whiting only. Chapter 13 includes the head muscles of the cod.

External features of the trout
Observe the stream-lined shape of the body and the arrangement of the fins. The caudal fin (tail) is moved in the horizontal plane by the body muscles and so propels the body forward; it is externally symmetrical (homocercal) but this symmetry does not extend to the skeleton. There are paired pectoral fins, situated ventrally just behind the head, and pelvic fins, also ventral and about half way along the body; these are used for balancing, turning, and braking. The unpaired median fins comprise two dorsals and one anal. The second (posterior) dorsal is small and fleshy and is called the adipose fin. All the other fins and the tail consist of bony fin rays supporting a tough membrane; muscles are confined to the base of the fins.

The body is covered with small round (cycloid) scales that

lie under the epidermis. The surface of the body is smooth and is covered in life with mucus. The body colour is variable: in brown trout there are usually red and black spots and a red border to the adipose fin, whereas rainbow trout lack this red border and have dark spots on the tail and an iridescent magenta flush along their sides. A row of scales almost along the midline of the body on each side is pierced by the pores of the lateral line.

The mouth is terminal and is bordered by the premaxillae and maxillae, forming the upper jaws, and the dentaries, forming the lower jaws. All these bones bear teeth and there are also teeth on the palatines and vomer in the roof of the mouth. Inside the mouth are small flaps of skin, the maxilliary and mandibular valves; these allow the respiratory current of water to pass into the mouth but prevent flow outwards.

The eyes are large and have no eyelids but are covered with a layer of transparent integument. The nostrils are situated just in front of the eyes. There are two openings on either side and the upper, inner ones are guarded each by a flap valve. Push a probe into the lower opening and observe that it emerges through the upper one. Note that there is no connection between nostrils and mouth.

At the back of the head on each side is the large crescentic opening of the gill chamber. This is bounded behind by the pectoral girdle and in front by the operculum, a flap of skin supported by bones, which acts as a valve allowing the respiratory current to pass out behind it but preventing flow inwards. Ventrally, the operculum is continuous with a fold of skin, the branchiostegal membrane, which is supported by bony rays which articulate with the ceratohyal bone. The branchiostegal membrane of one side meets with that of the other in the mid-ventral line of the throat.

Lift up the operculum and find four gill arches each supporting two rows of gill filaments (making eight hemibranchs). The pseudobranch, which represents the pseudobranch of the spiracle in the skate, lies under the operculum dorsally and may have a secretory function. The fifth gill arch is bound by skin to the anterior border of the pectoral girdle and bears no gill filaments. There are thus five gill slits. Examine the gills and note that the primary lamellae of each hemibranch project freely from the supporting arch (and are not attached along their length to a septum as in the skate). Respiratory exchange takes place in the numerous secondary lamellae, set perpendicularly to the primary lamellae.

Immediately in front of the anal fin are two openings lying close together, the large anus anteriorly and the smaller urino-genital aperture posteriorly.

External features of the roach

The description of the trout applies to the roach with the following exceptions.

The body is deeper and less stream-lined. There is a single dorsal fin only. The scales are relatively larger and easily become detached because the epidermis is very thin. The colour is dark grey or green dorsally, shading to silver on the ventral surface. The fins vary in colour but are usually yellow or red and the iris of the eye is red.

The mouth is bordered by the premaxillae, which are protractile, and the dentaries. The maxillae lie wholly behind the premaxillae. There are not teeth on the jaws or in the mouth but pharyngeal teeth are present and will be seen later (p. 39).

External features of the whiting

The description of the trout applies to the whiting with the following exceptions.

There are three dorsal and two anal fins, all consisting of soft bony rays supporting webs of skin. The pectoral fins are set half way up the side of the body and the pelvic fins are anterior and ventral to the pectorals. The pelvic girdle is articulated to the pectoral girdle. This arrangement of paired fins allows for efficient braking without danger of vertical movement and it is characteristic of more advanced Teleosts. The colour is silvery but slightly darker dorsally and there is a black spot above the base of the pectoral fin.

The mouth is bordered by the premaxillae and the dentaries and all bear teeth. The maxillae are toothless and lie wholly behind the premaxillae, another character of advanced Teleosts.

The anus and urinogenital apertures lie in the mid ventral line an inch or two behind the pelvic fins.

General directions for dissection of the trout

Make a ventral incision slightly to the left of and anterior to the anus. Cut forward just dorsal to the left pelvic fin and continue in the mid-ventral line to the level of the pectoral fins. Take care not to damage the internal organs. Lay the fish on its right side and cut upwards (dorsally) immediately behind the left pectoral fin and then backwards, through the muscles and ribs, along the upper border of the body cavity. Then cut down to the anus and remove the flap of body wall. At the anterior dorsal end of the hole, cut away more body wall as far forwards as the posterior edge of the pectoral girdle and down to the pectoral fin. Cut this fin away near its base.

Contents of body cavity
(Fig. 10)

The nearly transparent air bladder lies dorsal to the transparent peritoneum forming the dorsal roof of the visceral cavity. The dark red kidney lies dorsal to the air bladder and can be seen through it. The gonads vary greatly in size with

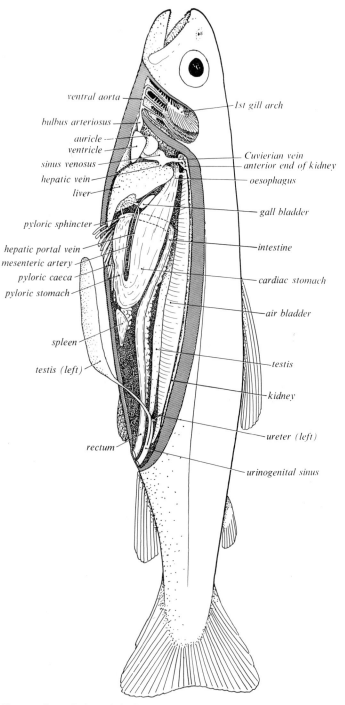

ventral aorta

bulbus arteriosus

auricle
ventricle
sinus venosus
hepatic vein
liver

pyloric sphincter

hepatic portal vein
mesenteric artery
pyloric caeca
pyloric stomach

spleen

testis (left)

rectum

1st gill arch

Cuvierian vein
anterior end of kidney
oesophagus

gall bladder

intestine

cardiac stomach

air bladder

testis

kidney

ureter (left)

urinogenital sinus

FIG. 10. Lateral view of the brown trout with the body wall, pectoral girdle, and operculum removed from the left side and the ventral aorta and afferent branchial arteries dissected out. The left testisac is pulled ventrally.

the age of the fish and season of the year. They may fill the body cavity of mature, nearly ripe, fish or be inconspicuous (see later for details).

ALIMENTARY CANAL. A wide oesophagus leads from the gill region (pharynx) into the muscular stomach. This is U-shaped and highly distensible. Its distal limb is separated by a pyloric sphincter from the intestine and there are a large number of finger-like pyloric caeca at the anterior end of the intestine. Trout from hatcheries usually have heavy fat deposits especially round these caeca. The intestine narrows slightly as it passes straight to the anus. The dark red spleen (which may be divided into several parts) lies between the tip of the stomach and the intestine. The pancreas is diffuse.

The liver is large and indistinctly lobed and the bile duct leads from it to open into the intestine among the pyloric caeca. The gall-bladder is a green sac embedded in the dorsal surface of the liver just below the oesophagus. Do not try to follow the bile duct back to the gall-bladder. Between the gall-bladder and oesophagus find the coeliaco-mesenteric artery which divides into branches supplying all the viscera. These are difficult to follow, as are the branches of the hepatic portal vein which can be seen entering the liver parallel with the bile duct.

URINOGENITAL SYSTEM. Carefully cut away the body wall on the left side posterior to the anus and find the median urinogenital sinus opening by an aperture behind the anus. In the *male* the testes are enclosed in a pair of testisacs which open into the urinogenital sinus. The testisacs extend forwards to the anterior end of the body cavity. In the mature fish they are large white or pinkish structures but in immature fish they are narrow almost colourless 'strings' immediately below the air bladder. There is no connection between testis and kidney. In the *female* the ovaries are not wholly enclosed in sacs but ripe eggs pass from the body cavity through funnels in the urinogenital sinus. Ripe ovaries distend the body cavity. In immature or resting fish, the orange ovaries lie at the front end of the body cavity immediately below the air bladder and are widest anteriorly, tapering posteriorly. In both sexes there are two ureters, one leaving the kidney on each side, and they join to form a muscular tube which opens into the urinogenital sinus. Carefully dissect the posterior end of the swim bladder away from the kidney and so expose the ureters leaving the kidney and joining each other ventral to the air bladder.

General dissection of the roach

Follow the instructions for the trout.

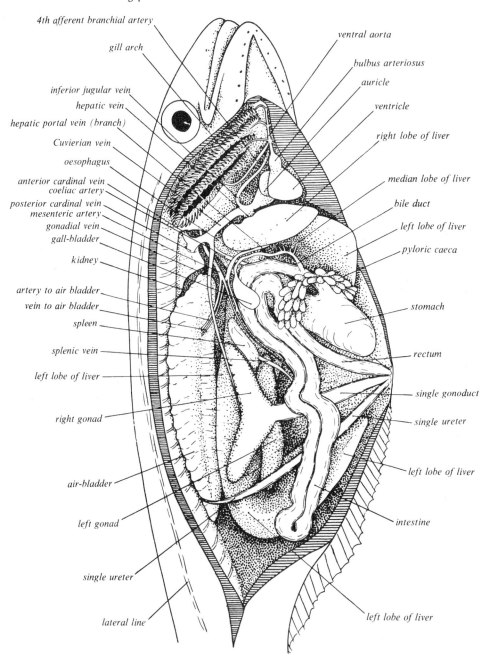

4th afferent branchial artery

gill arch

inferior jugular vein

hepatic vein

hepatic portal vein (branch)

Cuvierian vein

oesophagus

anterior cardinal vein
coeliac artery

posterior cardinal vein
mesenteric artery

gonadial vein

gall-bladder

kidney

artery to air bladder

vein to air bladder

spleen

splenic vein

left lobe of liver

right gonad

air-bladder

left gonad

single ureter

lateral line

ventral aorta

bulbus arteriosus

auricle

ventricle

right lobe of liver

median lobe of liver

bile duct

left lobe of liver

pyloric caeca

stomach

rectum

single gonoduct

single ureter

left lobe of liver

intestine

left lobe of liver

FIG. 11. Oblique ventro-lateral view of the whiting with the body wall, pectoral girdle, and operculum removed from the right side and the ventral aorta and afferent branchial vessels dissected out.

The air bladder has a thick white wall and lies dorsal to the transparent peritoneum forming the dorsal roof of the abdominal cavity; it is divided by a constriction into a smaller anterior part connected with the inner ear via Weber's ossicles (see later, p. 40) and a larger posterior part used in control of specific gravity. The dark red kidney lies dorsal to the air bladder and is visible at the constriction. The gonads vary greatly in size with the age of the fish and season of the year.

ALIMENTARY CANAL. Cyprinid fishes such as the roach have no stomach. The oesophagus leads straight into the intestine and this is long and coiled. There are no pyloric caeca. There may be thick fat deposits along the gut. The liver consists of small left and large right lobes and the gall-bladder is embedded in the central part. The bile-duct leads from the bladder to the beginning of the second loop of the intestine. Roach are omnivorous, including plant material, animals, and mud in their diet. Lift up the gonad and find the pneumatic duct entering the air bladder just posterior to the constriction. Follow it back to its origin from the dorsal surface of the oesophagus. Trace the branches of the coeliacomesenteric artery, found close to the oesophagus, and note the branches of the hepatic portal system lying parallel to the arteries but returning blood via the hepatic portal vein to the liver. The spleen may be divided into several parts which lie within the first loop of the intestine.

URINOGENITAL SYSTEM. Carefully cut away the body wall on the left side posterior to the anus and find the median urinogenital sinus opening by an aperture behind the anus. The gonads and kidneys have the same form in both sexes but ovaries usually obviously contain eggs even in immature fish. The gonads are enclosed in sacs that are continuous with the gonoducts which join just anterior to the urinogenital sinus. One ureter leaves the kidney on each side and the two join to form a muscular tube which opens into the urinogenital sinus.

Carefully dissect the posterior end of the air bladder away from the kidney to expose the ureters. Try not to puncture the air bladder.

General dissection of the whiting

Make a median ventral incision from the pelvic fins backwards to the posterior border of the body cavity. In carrying the cut backwards pass to the right of the anus, being careful not to injure the rectum, ureter, or gonoducts. Now cut upwards from the pelvic fins and remove the wall of the body cavity on the right side.

Note that the body cavity is lined with a dark pigmented peritoneum.

LIVER AND GALL-BLADDER. The liver is a large yellowish organ with the anterior end attached to the anterior boundary of the body cavity. It is divided into three lobes of which the left lobe is very much longer than the other two. The left lobe extends backwards as far as the posterior end of the body cavity.

The gall-bladder is a large greenish sac which is not embedded in the liver. It lies behind the short right lobe of the liver on the right side of the oesophagus. The cystic duct leaves the gall-bladder from its anterior end.

ALIMENTARY CANAL. The stomach is a large thick-walled tube. It is U-shaped and varies enormously in size according to the amount of food which it contains. The basal portion of the U of the stomach forms a large sac, the fundus, which is easily distensible in the presence of food. When it is completely empty it is no more than $\frac{1}{4}$ to $\frac{1}{2}$ inch long but when distended with food it extends backwards posteriorly to the hinder end of the body cavity. The distal limb of the stomach is separated from the intestine by the pyloric sphincter. The position of the pylorus is marked by the presence of numerous pyloric caeca. These are blind tubes communicating with the gut.

Behind the pyloric caeca the intestine curves posteriorly and passes close to the gall-bladder. Trace the cystic duct forwards from the gall-bladder. In about $\frac{3}{4}$ inch the cystic duct receives several hepatic ducts from the right lobe of the liver. The common bile-duct continues from this point and opens into the intestine just behind the pyloric caeca. The intestine after receiving the bile-duct passes backwards to the hind end of the body cavity, then loops forwards on itself and finally passes ventrally to the anus. The pancreas is a diffuse gland that is not easily seen in dissection. The pancreatic duct opens into the bile-duct.

Cut through the body wall on the left-hand side round the anus and urinogenital openings, leaving a portion of the body wall attached to these openings. Now take hold of the portion of body wall attached to the anus and urinogenital opening and turn it over to the right-hand side.

URINOGENITAL SYSTEM. Stretch the median mesentery connecting the urinogenital opening and the anus to the viscera. In this mesentery note the wide and broad ureter behind, the single gonoduct in the middle and the rectum in front. Follow the gonoduct running in the mesentery to its connection with the paired gonad. The gonads are elongated structures lying side by side and connected in the middle at the point where the gonoduct leaves them. The form of the gonads and the duct in the two sexes is identical. The ovaries and testes cannot be distinguished externally except in the breeding season when ripe. Note that the gonad is not connected with the kidney, nor, in the case of the female, is the oviduct open to the body cavity.

Trace the ureter dorsally and posteriorly. It reaches the dorsal side of the air bladder and then disappears from view on one or other side of the posterior end of the air bladder. (For further dissection see p. 47.)

THE AIR BLADDER AND SPLEEN. The air bladder lies in the middle line on the dorsal side of the body cavity. It extends almost the whole length of the body cavity. Only the ventral wall of the air bladder is seen in the inspection of the body cavity contents. This wall is tough and is covered with peritoneum. The dorsal wall is closely applied to the kidneys.

The spleen is an oval reddish body lying in the mesentery between the intestine and the anterior end of the gonads.

The vascular system of the trout (Fig. 12)

Continue the cut along the midline forwards cautiously to expose the heart in the pericardium. Carefully remove the ventral muscles and the left pectoral fin and girdle. Remove the left operculum and branchiostegal membrane by cutting the muscles at the upper edge of the operculum, dislocating it and then cutting along the line of bending. Roll up a piece of paper and insert it through the mouth into the pharynx so as to dilate the gill region.

Arrange the fish so that its ventral surface is upwards.

Dissect the connective tissue off the heart and identify the sinus venosus and the Cuvierian veins joining it. Follow the left Cuvierian vein dorsally and note the anterior (glandular) end of the kidney. The Cuvierian vein is formed by the joining of the anterior and posterior cardinal veins which will be seen later.

Cut forwards at the level of the ventricle to expose the ventral aorta. There are four pairs of afferent branchial arteries. The third and fourth leave the aorta opposite the base of the fourth gill arch; the third passes directly outwards to the base of the third branchial arch and the fourth runs obliquely backwards and dorsal to a strip of muscle to enter the fourth branchial arch. The second afferent branchial runs forwards and outwards to the second branchial arch and the first pair of afferent branchials are formed by the bifurcation of the ventral aorta near the bases of the first branchial arch. In the rainbow trout, the second, third and fourth pairs of afferent branchial arteries leave the ventral aorta very close together but in the brown trout the second arises more anteriorly than the other two pairs. Clear lamellar tissue and muscles to expose the arteries as they pass along the gill arches.

Remove the heart from the pericardial cavity and open it by a median longitudinal section and examine it under water in a small dish. Observe that the walls of the ventricle are very thick and muscular while those of the sinus venosus and auricle are thin. A thick-walled bulbus arteriosus leads from the ventricle and represents the root of the ventral aorta.

The following valves are present: the sinu-auricular valve, between the sinus venosus and the auricle (formed by two membraneous flaps anterior and posterior in position); the auriculo-ventricular valve, formed by two flaps, dorsal and ventral in position; the aortic valves, a pair of semilunar valves situated on the right and left sides at the junction between ventricle and bulbus arteriosus.

THE EFFERENT BRANCHIAL SYSTEM. Cut through the left angle of the jaw with scissors, carry the cut backwards through the branchial arches horizontally and then reflect the lower jaw and pin it back so as to expose the dorsal wall of the mouth and pharynx. Cut through the thick mucous membrane in the midline between the gill slits and pull it carefully aside to expose the efferent vessels at the dorsal ends of the gill slits.

The two anterior efferent branchial arteries on each side join paired lateral dorsal aortae which form part of a circulus cephalicus. They fuse anteriorly dorsal to the parasphenoid bone and join posteriorly to form the dorsal aorta. The two posterior efferent branchial arteries on each side join just before they enter the dorsal aorta. This then gives off a pair of subclavian arteries to the pectoral fins and a single large coeliaco-mesenteric artery which passes through the kidney and past the anterior end of the air bladder to supply the organs in the body cavity (see p. 33). To see this, cut the Cuvierian vein, noting the anterior and posterior cardinal veins forming it, and clear the connective tissue from the anterior end of the kidney and the air bladder. Find the pneumatic duct connecting the oesophagus dorsally with the air bladder. Regulation of the amount of gas in the air bladder occurs through this open duct.

The dorsal aorta continues down the midline of the body dorsal to the kidney and ventral to the vertebral column. It gives off renal arteries and segmental vessels to the body muscles.

The vascular system of the roach

Follow the instructions for the trout but note that the pectoral girdle has a broad inturned flange anteriorly. Cut through this girdle near its dorsal end and then pull it forwards, dissecting it away from the body.

THE AFFERENT BRANCHIAL SYSTEM AND HEART. These are very similar to those of the brown trout, see pp. 37–8.

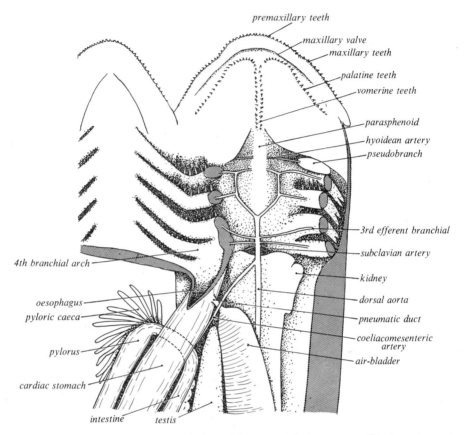

FIG. 12. Efferent branchial arterial system of the brown trout. The lower jaw and ventral parts of the gill arches have been reflected to the animal's right and the body cavity opened up. The mucous membrane on the roof of the pharynx has been removed and the blood-vessels dissected out. The air bladder has been pulled away from the kidney and the liver has been removed.

The pharyngeal teeth of the roach

In Cyprinid fishes, the ceratobranchial bones of the fifth branchial arch bear teeth on their buccal surfaces. These bite against a pad on the dorsal surface of the pharynx (see p. 40); they are shed and replaced through life and their number, shape, and arrangement are diagnostic of the species.

Cut through the left angle of the jaw and carry the cut backwards through the four branchial arches that bear gill filaments. The ceratobranchial bone of the fifth arch is very large. Remove it from the left side by cutting through the arch dorsal to it and then through the muscles posterior to it in the midline which hold it to the bone on the other side. Note the tips of the teeth visible through the flesh. There may be loose teeth present in the process of being shed. Drop the fifth ceratobranchial (also called the pharyngeal bone) into a beaker of boiling water (or *dilute* caustic soda). When the flesh is softened, tease it away to expose the bones and the teeth. Clean by immersing in hydrogen peroxide and removing all trace of soft parts with forceps. Draw the 'pharyngeal bone', showing the position and shape of the teeth. In the roach, there should be five or six teeth in one row on each bone.

THE EFFERENT BRANCHIAL SYSTEM. Reflect the lower jaw and pin it back to expose the dorsal wall of the mouth and pharynx. Note in the midline behind the gill slits the bony pad against which the pharyngeal teeth bite. Remove this by cutting round it with a scalpel and levering it away and note that it leaves a socket in the 'suprapharyngeal' bone which is firmly fused with the parasphenoid bone of the skull and has a posterior flange from which muscles arise.

Cut through the mucus membrane in the midline anterior to this bone and pull the membrane aside carefully to expose the efferent vessels at the dorsal ends of the gill slits. The pattern of these vessels is identical with that in the trout, so see Fig. 12. The two anterior arteries on each side join the circulus cephalicus. The two posterior arteries of each side join each other and the median dorsal aorta immediately in front of the suprapharyngeal bone and the subclavian arteries leave the aorta here and proceed to the pectoral fins.

Posterior to the suprapharyngeal bone there is a strong ventral vertebral process to which the anterior end of the air bladder and the anterior end of the peritoneal lining are attached. The head kidney bulges on either side. Find the anterior and posterior cardinal veins joining to form the Cuvierian veins. Slice horizontally dorsal to the suprapharyngeal bone and expose the dorsal aorta in the midline. Find the coeliaco-mesenteric artery leaving it on the right side and passing through the kidney tissue to enter the visceral cavity to supply the gut and other abdominal organs.

Weber's ossicles in the roach (Fig. 13)

These are a series of small bones connecting the tunica externa of the anterior end of the air bladder on each side with a median cavity containing perilymph. Sound waves cause changes in the volume of gas in the anterior part of the air bladder and these changes cause movements of the edge of a slit in the tunica externa, which can slip over the intact tunica interna. The tripus is attached to the edge of this slit and also pivots on the third vertebra. The movements of the slit are transmitted via the tripus to the smaller ossicles anterior to it and so to the perilymph. Movements of this cause movements of the endolymph in a canal joining the sacculi of the two inner ears and the consequent vibration of the saccular otoliths results in hearing. The Ostariophysi, all of which have Weber's ossicles in some form, are noted for their acute sense of hearing.

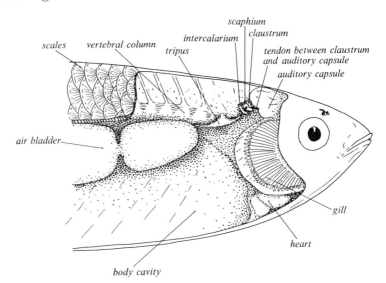

FIG. 13. Roach with the right body-wall and muscles cut away exposing the Weberian ossicles and the air bladder. The tendons between the ossicles are drawn in black.

If the efferent arterial system has been dissected, proceed as follows. Carefully remove the muscles at the anterior end of the air bladder below the ribs and so find the tripus, attached posteriorly to the tunica externa of the air bladder and running anterodorsally to disappear ventral to the first rib. Move the rounded outer edge of this bone and note that it pivots easily and stretches the air bladder. Remove the skin and the epiaxial (dorsal myotomic) muscles to reveal the dorsal surface of the tripus. Use a lens or dissecting microscope to find the other ossicles. These are visible to the naked eye only as dark interruptions in the shining white ligament which is attached to the front end of the tripus and proceeds anteromedially to end on the basioccipital bone. The small claustrum touches a membraneous window in this bone with the sinus impar, containing perilymph, on the other side. The scaphium is very close to it and the intercalarium lies midway between tripus and scaphium and is articulated to the second vertebra. It is possible to remove the chain of ossicles held together by the ligaments and to examine them in a watch-glass.

If the roach is to be dissected only to demonstrate Weber's ossicles, proceed as follows. Remove the lower part of the abdominal wall of the roach and note the air bladder. Remove the muscles from the outer side of the ribs and vertebrae in the anterior part of the body right up to the skull. The first rib ,behind the skull can be distinguished by its having a long outer process and a shorter inner one. The inner and shorter one lies slightly deeper in the body wall and extends obliquely forwards. Remove the outer process of the first rib. Cut through all the ribs posterior to the first one close to the vertebral column and remove them, so exposing the air bladder.

Just within the inner process of the first rib find the outer edge of the most posterior ossicle, the tripus. Carefully cut through the inner process of the first rib close to the vertebral column. Remove this inner process taking care not to disturb the tripus. Uncover the whole of the tripus; posteriorly it is attached to the anterior end of the swim bladder and at its anterior end a white tendon connects it to the next ossicle, the intercalarium. For the dissection of the ossicles anterior to the tripus, it will be necessary to use a lens. The intercalarium is small and extends dorsally as a thin plate. It is connected by a tendon to the next ossicle, the cup-shaped scaphium. The claustrum is anterior to, and is covered by, the scaphium. The claustrum and the scaphium are so close together that they appear, unless carefully dissected apart, to be one ossicle. Finally find the tendon connecting the scaphium to the posterior border of the auditory capsule.

The vascular system of the whiting
(Fig. 11)

General directions

Continue the median ventral incision forwards as far as the branchiostegal membrane. In doing this the pericardial cavity is opened. Observe the heart lying in the pericardial cavity. Carefully cut away the right-hand side of the pectoral girdle lying along the sides of the pericardial cavity and along the posterior border of the gill chamber. Return the viscera to the left-hand side of the body.

Arteries and veins of the body cavity

Along the dorsal border of the gonad find the spermatic (or ovarian) vein. Follow this vein forward to its junction with the large Cuvierian vein, which has been exposed by the removal of the pectoral girdle, and lies between the right lobe of the liver and the posterior border of the gill chamber. Follow the Cuvierian vein ventrally to its junction with the sinus venosus. In passing the right lobe of the liver the Cuvierian vein receives one of the paired hepatic veins. Follow the Cuvierian vein dorsally to the point where it is formed by the union of the posterior and anterior cardinal veins. The end of the posterior cardinal vein will be seen emerging through the kidney. The anterior cardinal vein runs forwards dorsal to the gill arches and a short posterior portion of it can be seen joining with the posterior cardinal.

Above and on the right-hand side of the oesophagus will be seen two arteries, the anterior of the two is the coeliac and the posterior the mesenteric artery. A branch of the mesenteric artery opposite the anterior end of the air bladder goes to the air bladder and penetrates the ventral wall of this organ in the middle line a short distance from its anterior end. The mesenteric artery continues posteriorly along the wall of the intestine, spleen, and gonads. Running parallel with the mesenteric artery lies a large branch of the hepatic portal vein. It receives a large branch draining the spleen and a vein from the air bladder lying parallel with the artery of this organ. Anteriorly the hepatic portal vein curves round parallel with the bile-duct and enters the right lobe of the liver.

The pericardial region

The walls of the pericardium often remain closely applied to the heart after the pericardial cavity has been opened. If this is so, remove the thin membrane surrounding the heart. The ventricle is a large triangular thick-walled organ occupying the major portion of the pericardium as in the skate. A single thin-walled auricle lies dorsal to the ventricle and can be seen in ventral view projecting on either side of the ventricle. The sinus venosus is a transverse triangular sac lying across the posterior part of the pericardium. It receives laterally the Cuvierian veins and opens into the auricle in the middle line. Just before the Cuvierian vein enters the sinus venosus it receives the inferior jugular vein on the right side of the body. Follow this vein forwards where it is visible through the thin roof of the pericardium. It drains the ventral pharyngeal region in the middle line. Usually this vein is fully developed on the right-hand side only.

The afferent branchial system

There is no conus arteriosus visible as in the skate. The ventral aorta leaves the anterior end of the ventricle and dilates at its root to form the muscular bulbus arteriosus. From the bulbus arteriosus follow the ventral aorta forwards. It is about 1 inch in length. Opposite the base of the fourth gill arch the third and fourth afferent branchial arteries leave the ventral aorta. The fourth afferent branchial arises just behind the third and runs obliquely backwards and upwards dorsal to the inferior jugular vein on the right side. It passes dorsal to a strip of muscle and enters the fourth branchial arch about half an inch from its lower end. The third afferent branchial artery passes directly outwards to the base of the third branchial arch. The second afferent branchial leaves the ventral aorta about $\frac{1}{8}$ inch in front of the third and runs forwards and outwards to the second branchial arch. The ventral aorta bifurcates anteriorly to form the first afferent branchials which pass forwards and outwards to the bases of the first branchial arches.

Dissection of the heart

Remove the heart from the pericardial cavity by cutting through the ventral aorta just in front of the bulbus arteriosus and through the Cuvierian veins. Open it by a median longitudinal incision made with the scissors. The examination is best done under water in a small dish.

Observe that the walls of the ventricle are very thick and muscular. The walls of the auricle and sinus venosus are thin, while the walls of the bulbus arteriosus are thick and fibrous. The valves are (1) the sinu-auricular valve between the sinus venosus and the auricle, formed by two membranous flaps, respectively anterior and posterior in position; (2) the auriculo-ventricular valve formed by two flaps, respectively dorsal and ventral in position; (3) the aortic valves, a pair of semilunar valves situated on the right and left sides at the junction between the bulbus arteriosus and the ventricle.

The efferent branchial system (Fig. 14)

Remove the ventral wall of the mouth and pharynx by cutting through the angles of the jaw with a strong pair of scissors. Carry the cut backwards through the operculum and the branchial arches in a horizontal direction finishing at the posterior border of the operculum. On the roof of the pharynx between the cut gill arches will be seen a pair of round thickenings in the mucous membrane bearing numerous teeth pointing backwards. These pharyngeal teeth are fused to a bony plate underlying the mucous membrane and formed by the fusion of the pharyngo-branchial bones. Now cut through the mucous membrane covering the roof of the mouth in the middle line. Carry this cut backwards behind the pharyngeal teeth. Carefully lift up one flap of the mucous membrane and the dorsal aorta will be visible in the middle line just behind the posterior border of the pharyngeal teeth. In the whiting the two dorsal aortae do not unite posteriorly to form the median dorsal aorta until they reach the anterior end of the body cavity. The paired aortae are united also anteriorly near

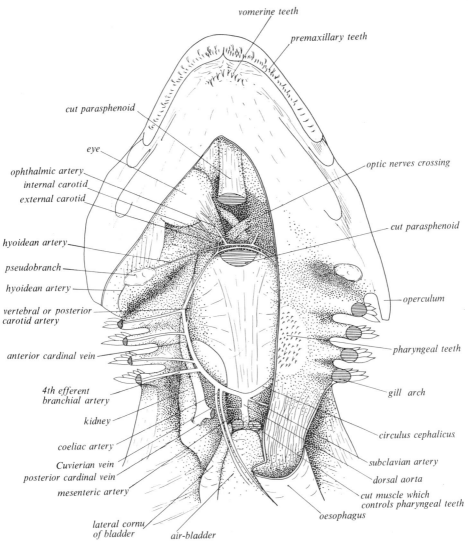

vomerine teeth

premaxillary teeth

cut parasphenoid

eye

ophthalmic artery
internal carotid
external carotid

optic nerves crossing

cut parasphenoid

hyoidean artery

pseudobranch

hyoidean artery

vertebral or posterior
carotid artery

operculum

anterior cardinal vein

pharyngeal teeth

4th efferent
branchial artery

gill arch

kidney

coeliac artery

circulus cephalicus

Cuvierian vein
posterior cardinal vein

subclavian artery

mesenteric artery

dorsal aorta

cut muscle which
controls pharyngeal teeth

lateral cornu
of bladder

oesophagus

air-bladder

FIG. 14. Efferent branchial arterial system of the whiting. The lower jaw and the ventral parts of the gill arches have been removed. The mucous membrane on the roof of the pharynx and mouth has been cut in the middle line and removed on the right side.

the internal carotid arteries and form a ring, the circulus cephalicus (see Fig. 14). Raise one half of the mucous membrane on the right side and follow one of the paired aortae forwards and outwards from the dorsal aorta. Follow round

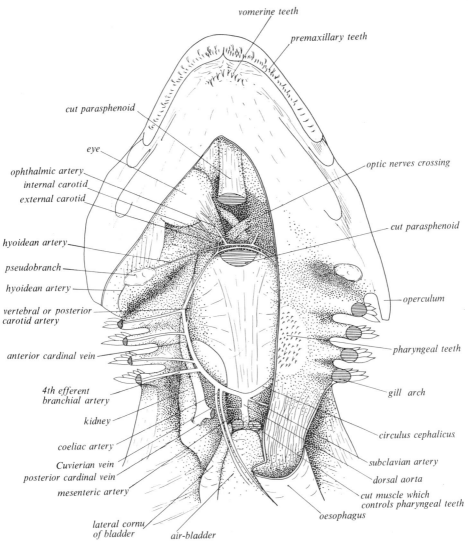

FIG. 14. Efferent branchial arterial system of the whiting. The lower jaw and the ventral parts of the gill arches have been removed. The mucous membrane on the roof of the pharynx and mouth has been cut in the middle line and removed on the right side.

the internal carotid arteries and form a ring, the circulus cephalicus (see Fig. 14). Raise one half of the mucous membrane on the right side and follow one of the paired aortae forwards and outwards from the dorsal aorta. Follow round

one half of the ring and note the vessels which are given off
from it. Very close to the dorsal aorta there arise two vessels
side by side. The outer of these is the coeliac artery and the
inner the mesenteric artery. Trace these vessels backwards to
the viscera. These median arteries usually arise from the dorsal
aorta. Here, however, the paired dorsal aortae do not unite
until after the origin of these arteries, which are thus left
arising from one of the paired dorsal aortae.

Quite close to the origin of the mesenteric artery will be
seen the origins of the fourth and third efferent vessels. A little
farther forward will be found the separate origins of the
second and first efferents.

After finding the origin of the efferent branchials remove
the mucous membrane and the pharyngeal teeth from the roof
of the mouth and over the gill arches. Trace the efferent
branchial vessels outwards to the gill arches which they supply. Between the origin of the first and second efferent branchial vessels observe the vertebral (or posterior carotid) artery
arising from the inner side of the aorta. It runs forwards and
passes into the skull.

Consult the diagram (Fig. 14) and note that the aorta continues forwards after the origin of the first efferent branchial
and meets its fellow on the opposite side. Just before this
union the paired external and internal carotids are given off.
In the same diagram note the hyoidean artery arising directly
from the first efferent branchial and running to the pseudobranch. It breaks into capillaries in the pseudobranch and
emerges on the inner side of this organ, whence it runs inwards to the middle line to meet its fellow on the opposite side
immediately in front of the union of the dorsal aortae. Just
before this union, the hyoidean gives off a small branch, the
ophthalmic artery, which runs to the eyeball.

The union of the aortae and the hyoidean vessels cannot be
seen without cutting through the parasphenoid bone in which
they lie. Follow the aortae inwards until they disappear, and
then cut through the parasphenoid bone carefully at this point
and try to trace the vessels as shown in the diagram.

The paired subclavian arteries are not quite constant in
place of origin. They arise on either side from the paired
dorsal aortae just as these unite to form the median dorsal
aorta, or they are given off from the anterior end of the median
dorsal aorta itself. The subclavian arteries pass outwards
dorsal to the kidneys to supply the pectoral fins.

The anterior end of the kidney forms a large red (brown in
preserved specimens) body lying dorsal to the posterior parts
of the paired aortae and the third and fourth efferent branchials. Pick up the cut end of the Cuvierian vein lying over

the kidney. From it trace the anterior cardinal vein forwards through the kidney immediately dorsal to the paired aortae and branchial vessels.

The kidney and the air bladder

The position of the head of the kidney has already been described. On the inner side of the head of the kidney observe the large muscle which works the pharyngo-branchial teeth. Carefully dissect away the peritoneum from the anterior end of the air bladder and note that the anterior end of this organ is divided into a middle portion with two smaller cornua on either side. These cornua cover over the kidney just behind the head portion. Raise the cornu on one side, and observe the kidney extending backwards.

Cut open the air bladder by a longitudinal incision slightly to one side of the middle line. Turn back the ventral wall of the swim bladder and note the large vascular plexus. Gently turn the tissues of the vascular plexus away from the ventral wall of the swim bladder and notice the artery and vein already described, penetrating through the wall of the swim bladder and supplying all parts of the vascular plexus. This is the gas gland from which oxygen is secreted into the cavity of the air bladder.

The kidney must now be exposed. In order to do this remove the air bladder covering it in the anterior region; and in the posterior region, strip away the thin roof of the air bladder from the kidneys which lie dorsal to it. In doing this part of the posterior cardinal veins may be torn open. At the posterior ends of the kidneys find two small ureters leaving them and joining together just anterior to the place where the common ureter passes out round the side of the swim bladder. If possible trace the two ureters forwards along each kidney. Behind the kidneys find the caudal artery and vein, the latter divides into two renal portal veins.

The central nervous system (Fig. 15)

There is a common plan for the brain and the cranial and spinal nerves of all Teleost fishes but different senses are emphasized in different species. This is accompanied by differences in the relative sizes of the parts of the brain and the presence or absence of special sensory nerves. In the trout sight is the most important sense, while the roach probably also has good hearing (see above). Both these species lack the special cutaneous nerves (branches of VII and X) that are found in the whiting.

For most species of Teleost fish, the account of the whiting which follows will provide an adequate guide for dissection if the first section is omitted. Start at 'Exposure of the Brain'

and be prepared for difference in proportions and some differences in detail.

**The central
nervous system
of the whiting**
(Fig. 15)
*The sub-cutaneous
sensory and
lateralis nerves*

Immediately below the skin in the region of the operculum,
and along its upper border, lie a number of nerves. These
nerves belong to the lateralis and viscero-sensory system.

Carefully remove the skin from the left side of the body
behind the head. Immediately above the upper border of the
operculum will be found the cutaneous nerve, which arises in
the brain from the same roots as the VII and X cranial nerves.
Follow this nerve forwards through the muscles over the back
part of the skull to its origin through the roof of the skull
opposite the junction between the cerebellum and the optic
lobes. About $\frac{1}{2}$ inch behind its origin this nerve gives off a
small branch from its upper border which passes to the base
of the dorsal fin and runs backwards to one side of the mid-
dorsal line. This nerve is very delicate and is very often
severed in removing the skin. Follow the main branch of the
cutaneous nerve obliquely backwards along the posterior
border of the gill chamber. It passes under the upper lateral
line nerve, the position of which is easily determined by the
external marking. After passing under the upper lateral line
nerve it goes downwards along the border of the gill chamber
to the pectoral fin. In its course downwards it passes over the
lower lateral line nerve. Follow the cutaneous nerve downwards round the posterior border of the gill chamber; just
before reaching the upper edge of the fin it divides into two.
One branch runs ventrally, following the posterior border of
the gill chamber to the bases of the pectoral and pelvic fins.
The other branch runs backwards and crosses over the lower
lateral line nerve where it again divides. Of these, one branch
runs ventrally behind the fin to its base, and the other branch
runs backward obliquely and reaches the ventral side of the
body near the anal fin. The cutaneous branches of the nerve
end in sense organs of a viscero-sensory nature.

Trace the two branches of the lateral line nerves backwards.
It will be found that the upper one runs underneath the externally marked lateral line, while the lower one runs in a line
formed by the lower angle of the V-shaped myotomes.

*Exposure of the
brain*

Strip off the skin and muscle from the roof of the skull. Carefully slice through the frontal bone so as to expose the ophthalmic nerve (V and VII) running below it. Trace this nerve

forwards where it divides into many branches supplying the skin of the snout. It is not possible to distinguish the lateralis from the somatic sensory branches of this nerve. Follow the ophthalmic nerve backwards to its emergence from the orbit. Now slice away the bony roof of the skull exposing the brain both behind and between the eyes, but being careful not to injure the root of the cutaneous nerve. Note that the major portion of the brain lies behind and not between the eyes as in skate. The olfactory lobes will be found underneath the exposed portion of the ophthalmic nerve. They are connected to the rest of the forebrain lying behind the eyes by a pair of long slender olfactory tracts.

Nerves supplying the eye and its muscles, and nerves lying across the floor of the orbit

Remove the skin carefully from over one eye and cut away the bony roof of the orbit so as to expose the ophthalmic nerve lying across the eye muscles. Along the posterior border of the orbit a small nerve will be seen passing across the muscle in this region and ending near the skin along the lower border of the orbit. This is the outer buccal nerve (VII), which does not run across the floor of the orbit as in skate.

Now cut through the bony wall of the cranium so as to expose the origin of the ophthalmic nerve from the brain. Note that there is no profundus branch of the ophthalmic nerve in the whiting.

The arrangement of the eye muscles and their nerves is essentially similar to that in the skate. There is no bony inter-orbital septum and the pair of obliquus muscles arise anteriorly, close to the pair of the other side. Similarly, the two quartets of rectus muscles arise close together under the neurocranium, usually in a special skeletal recess called the posterior myodome. Gadiform fish such as the whiting lack myodomes but these are present in the trout and roach. The III, IV, and VI nerves enter the orbit together from its posterior angle. In following the III, IV, and VI nerves notice the optic nerve passing to the eyeball. In order to see the optic nerve inside the skull, very carefully raise the forebrain.

Now remove the eyeball by cutting through its muscles and nerves close to the cranium. From the posterior inner angle of the orbit will be seen a band of nerves below the outer buccal passing downwards and outwards. Half way across the orbit the greater part of this band of nerves passes out of view below the floor of the orbit.

This band of nerves comprises the inner buccal and the maxillary and mandibular branches of the V nerve, which can just be seen lying side by side. The outer nerve is the mandibular (V). Trace this nerve outwards towards the angle of the jaw, removing the *adductor mandibulae* muscle where

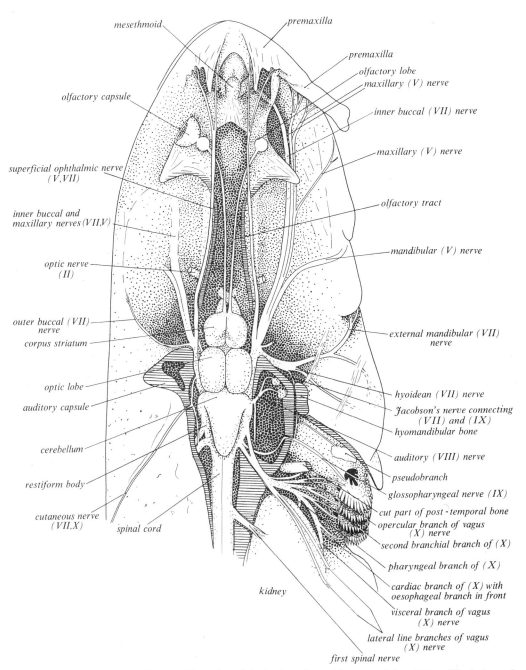

FIG. 15. Dissection of the brain and cranial nerves of the whiting. The lachrymal (nasal) bones have been carefully removed and the roof of the brain case cut away. Both eyes have been removed and also the connective tissue covering the floor of the right orbit. On the right side the dissection has been carried deeper down to expose the vagus nerve. The cut bone-surfaces are cross-hatched.

necessary and noting its nerve supply. V then curls inwards through the bone in this region to supply the lower jaw. Follow the remaining nerves forwards across the floor of the orbit. When they reach its anterior edge the inner buccal separates from the maxillary nerve and passes inwards and downwards to supply the lateral line sense organs of the snout. The maxillary nerve (V) after separating from the inner buccal sends numerous sensory branches to the skin over the premaxilla and maxilla bones.

Dissection of the VII nerve

The inner and outer buccal and the ophthalmic branches of the VII nerve have already been seen. Carefully remove the *levator hyoidei* muscle lying behind the outer buccal nerve so as to expose the anterior outer face of the auditory capsule. Immediately below the origin of the outer buccal and slightly behind it can be seen the large hyomandibular nerve emerging through the wall of the cranium. Follow this nerve outwards towards the operculum removing the muscles over the opercular bones. In about $\frac{1}{4}$ inch, just before reaching the hyomandibular bone, the hyomandibular nerve divides into two branches. The anterior branch is the external mandibular and passes through a flange from the hyomandibular bone and runs downwards along the front border of the operculum. Trace it forwards to the angle of the jaw where it breaks up to supply a group of lateral line ampullae. Shortly after passing through the hyomandibular bone, the external mandibular nerve gives a small branch, which passes downwards and backwards through one of the opercular bones to supply a group of lateral line organs situated between these bones.

Now trace the posterior branch (hyoidean) of the hyomandibular nerve. The hyoidean nerve supplies the *levator hyoidei* and opercular muscles and then passes through a separate foramen in the hyomandibular bone and runs backwards and downwards behind the external mandibular nerve, passing through the opercular bones. Follow it backwards between these bones until it passes into the branchiostegal membrane. You will find the nerve beneath the skin of the posterior side of the membrane lying along the lower border of the ceratohyal bone.

Notice that the ophthalmic, maxillary and mandibular V, inner and outer buccal and the hyomandibular VII nerves all arise from a common root. From the base of the hyomandibular nerve on its anterior side arises the palatine nerve. Trace the palatine nerve forwards. It lies near the middle line below the epithelium covering the floor of the orbit and supplies the roof of the mouth.

Follow the cutaneous nerve, which has already been

exposed, inwards through the cranium wall, removing the roof of the skull in this region if this has not already been done. Just within the cranium the cutaneous nerve is formed by the union of two branches, one from the base of the VII nerve and one from the posterior of the two roots of the X nerve.

Dissection of the
VIII nerve

Remove the top of the auditory capsule by an oblique cut backwards. If the upper border of the cut runs beside the cutaneous nerve already exposed, the result will be that the interior of the auditory capsule is exposed. Now remove the muscle behind the cutaneous nerve thus exposing the bone of the back of the skull and vertebral column. Continue to slice horizontally through the auditory capsule, so exposing the semicircular canals and membranous labyrinth. In the anterior part of the capsule note the branches from the VIII nerve supplying the ampullae of the semicircular canals as in the skate. Within the sacculus lies a long, bone-like concretion, the otolith. With a fine pair of forceps gently remove the otolith from the auditory capsule by drawing it forwards. In the posterior portion of the capsule the large band formed by the IX and X nerves passes across the capsule. The root of the VIII nerve can now be seen entering the anterior part of the capsule. From it a long branch passes backwards near the middle line to supply the utriculus and sacculus.

Dissection of the IX
and X nerves

The roots of the IX and X nerves have already been seen. Carefully uncover the IX and X nerves as they emerge from the outer side of the auditory capsule. The X nerve arises from the medulla by two roots which unite within the auditory capsule. Between these roots arises the IX nerve which passes outwards across the auditory capsule and penetrates its wall a short distance in front of the X nerve. Follow out the IX nerve. It divides into two on leaving the auditory capsule. One branch passes outwards to the first branchial arch and the other supplies the pseudobranch. From this second branch a small nerve (Jacobson's nerve) passes forwards close to the auditory capsule and unites with the hyomandibular nerve.

Follow the vagus (X) nerve through the posterior wall of the auditory capsule. Behind the capsule some of the branches of the vagus may be seen lying over the front part of the kidney. A large branch passes directly outwards in front of the kidney towards the operculum. This is the opercular branch and may have already been cut. The two large nerves running across the kidney are the lateral line branches of the vagus. Follow them backwards behind the gill region to the point where they have already been dissected out.

Gently raise the outer anterior edge of the kidney in order

to expose the remaining branches of the vagus. Close to the auditory capsule the first and second branchial branches are given off to the second and third branchial arches. The third branchial nerve leaves the vagus a short distance behind the others. These nerves divide and pass on either side of each branchial cleft. From the inner side of the vagus opposite the origin of the third branchial nerve arises the pharyngeal branch which curls inwards below the kidney to supply the pharynx. Close to the last gill arch, two branches leave the vagus close together and pass round to the ventral side immediately behind this arch. The anterior branch supplies the ventral wall of the oesophagus and the posterior branch (cardiac) goes to the heart. The main branch of the vagus forms the visceral nerve which passes directly backwards behind the origin of the cardiac and oesophageal nerves to supply the viscera, passing along the sides of the oesophagus into the body cavity.

Hypoglossal and first spinal nerves

A short distance behind the origin of the vagus nerve the first spinal nerve leaves the vertebral column. It passes backwards and outwards through the kidney and runs underneath the upper lateral line nerve to supply the pectoral fin. When the first spinal nerve reaches the pectoral fin the hypoglossal nerve separates from its anterior border. The hypoglossal nerve runs downwards along the posterior border of the gill chamber close to the pectoral girdle and Cuvierian vein. Ventrally it curls forwards to supply the hypobranchial musculature of the floor of the pharynx.

The brain
(Fig. 15)

The brain should now be examined.

DORSAL VIEW. The olfactory lobes lying against the nasal capsule, and the olfactory tracts have already been seen. The olfactory tracts lead back to two lobes which have the appearance of cerebral hemispheres. The lobes are really protrusions upwards formed by the thickened floor (corpora striata) of the forebrain. They are covered by a very thin transparent nonnervous roof, which is usually removed adhering to the skull with the membrane covering the brain.

The relationship of the Teleost forebrain with that of the Elasmobranch is seen in Fig. 9, b and c, p. 27.

Immediately behind the corpora striata lie the paired optic lobes. Between the optic lobes and the corpora striata can be seen a thin roof (the choroid plexus) covering over the third ventricle. Immediately behind the optic lobes lie the cerebellum above and medulla oblongata below. On either side of the cerebellum lie the two restiform bodies.

The posterior apex of the cerebellum covers over the roof

of the fourth ventricle, which can be seen by raising the cerebellum.

VENTRAL VIEW. Cut through the nerves as far from the brain as possible. Carefully remove the brain from the cranium.

At the anterior end observe the optic nerves crossing over but not fusing to form a chiasma. The ventral aspect of the brain does not differ further in any essential points from that of the skate which has already been described. Note the origin of the cranial nerves from the brain. The only marked differences from the skate are the positions of the origins of the IX and X nerves which have already been described.

Make a parasagittal cut through the brain. Note that the forebrain appears to be a solid structure. (Why?) Compare the structure of the forebrain of the whiting with that of the skate.

The eye

Remove the eye from the orbit. The relationships of the different layers will be found to be essentially the same as in the skate (p. 28).

4 *The salamander*

**External
features**

NOTE the general appearance of the animal. The head is large and indistinctly marked off from the body, there being no marked neck region. The limbs are short and project from the sides of the body. In the living animal the limbs are only capable of raising the body a very short distance off the ground. The skin is soft and slimy owing to the presence of glands. It plays a considerable part in respiration and is supplied with blood-vessels for this purpose. There are no scales, dermal ossicles, or fins.

In the male newt, *Triton cristatus*, there is a median dorsal crest and the tail bears a median dorsal and ventral fin, which is, however, unsupported by fin rays.

The vivid colour pattern of the salamander is an example of 'warning colouration'. The animal is provided with 'poison' glands the openings of which are clearly seen behind the eyes and along the back. These glands secrete a distasteful substance making the animal unpalatable, and their presence is correlated with the 'warning colouration'. Squeeze the glands and note the milky fluid that is expelled from them.

The mouth is terminal and is provided with small teeth. The tongue is firmly attached to the floor of the mouth and cannot be protruded as in the frog.

The external nostrils are very small and are situated on either side of the snout.

The eyes are large and on either side of the head. They are provided with movable eyelids in the salamander, but not in other *Urodela*. The tympanic membrane is absent.

The fore limb has four digits (thumb missing) and the hind limb has five digits.

The cloaca is a longitudinal slit in the middle line on the ventral surface just behind the hind limbs. It is provided with tumid lips.

The body narrows behind the cloaca to form the tail.
There are no external sexual differences in the salamander.

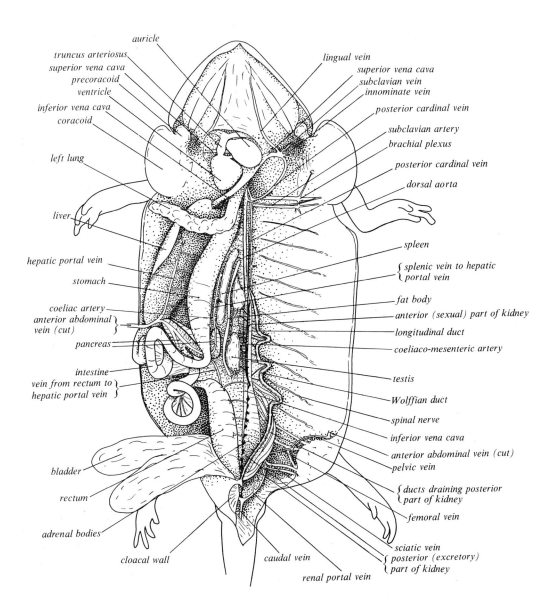

FIG. 16. View of the veins round the heart and of the contents of the body cavity of a male salamander. The pectoral girdle has been pulled apart, the pelvic girdle has been cut through and laid back, and the renal portal system dissected out.

General directions for dissection

Pin the animal on its back by a pin thrust through each limb. Cut through the skin in the mid-ventral line from the cloaca to the chin and pin it back on either side. Do not stretch the skin taut in the region of the arm-pits or you will tear the musculo-cutaneous vein, which should be looked for in this region.

Observe the anterior abdominal vein lying in the middle of the posterior portion of the abdominal wall. Unlike the frog it receives many large branches from the musculature of the abdominal wall. Make two small incisions through the abdominal wall on either side of the anterior abdominal vein. Ligature twice as near as possible to the liver and cut the vein between the ligatures. Carry one incision in the abdominal wall forwards to the pectoral girdle, being careful not to puncture the liver or the heart with the point of the scissors, and backwards to the pelvic girdle. Gently pull the two halves of the pectoral girdle apart, thus exposing the heart, and pin back the coracoids on either side. Notice the precoracoid processes by the pericardial membrane.

Anterior and middle regions of the body cavity (Figs. 16, 17)

Note the heart lying enclosed in its pericardium. On either side in the anterior part of the body cavity lie the black lungs. Between them lies the liver, which is a large flat brown body suspended by mesentery from the dorsal body wall. The inferior vena cava emerges from the anterior border of the liver, passes forwards in the falciform ligament and enters the pericardium to join with the superior venae cavae to form the sinus venosus. The gall-bladder may be seen by turning the posterior end of the liver forwards. It lies in a groove on the right side. Turn the liver over to the right-hand side as shown in Fig. 16, being careful not to destroy any mesenteries. The alimentary canal and its glands can now be traced.

The stomach is a wide tube lying parallel with the liver extending three-quarters of the way down the abdomen. At its posterior end it is constricted to form the pylorus. From the pylorus the intestine runs forwards towards the liver, forming the duodenum.

The duodenum and the stomach are attached to the liver by a mesentery.

The spleen lies to the right of the stomach. It is a large red body attached to the mesentery which suspends the stomach from the dorsal wall of the body cavity.

Between the duodenum and the stomach lies the pancreas. One lobe extends from the liver to the pyloric region, while the other lobe lies dorsal to the stomach and ends near the spleen. Gently squeeze the gall-bladder with a pair of forceps, and note the green bile passing down the bile-duct into the intestine at the distal end of the duodenum. The duct is short, not more than $\frac{1}{8}$ inch in length. Just before it enters the intestine it receives the ducts of the pancreas.

Cloacal region of the body cavity (Figs. 16, 17)

The intestine continues backwards and is coiled until it dilates to form the rectum. The rectum opens into the cloaca and thus passes to the exterior. In order to trace the rectum to the cloaca it is necessary to cut through the symphysis of the pelvic girdle in the middle line. Pull the two halves of the girdle gently apart and pin them out on either side. Be careful not to injure the large, bilobed, urinary bladder which opens into the cloaca immediately under the pelvic girdle. The kidney is a brown elongated oval body lying alongside the rectum attached to the dorsal body wall. Attached to the dorsal body wall on either side of the middle line towards the centre will be seen the gonads. Along the inner side of each gonad is a yellowish fat body, varying in size with the condition of the animal.

Arteries and veins of the body cavity (Figs. 16, 17)

Find the hepatic portal vein running through the pancreas to the dorsal side of the liver. Just before it enters the liver it receives the anterior abdominal vein. As it passes along the dorsal surface of the liver (in which it breaks up into capillaries), it receives branches from the stomach. As it passes through the pancreas it receives branches from the spleen, rectum, and small intestine.

Turn the viscera over to the left side of the animal and find the large inferior vena cava arising in the middle line between the two kidneys which it drains. At the level of the anterior end of the rectum it passes ventrally downwards to enter the lobe of the liver close to the gall-bladder. It runs through the liver and in its passage through this organ it receives the hepatic veins. It passes out of the liver at its anterior extremity, where it has already been noted.

The posterior cardinal veins are two small vessels that lie on either side of the dorsal aorta in the anterior part of the body cavity. Posteriorly they unite and join the inferior vena cava at the point where it leaves the dorsal body wall on its course to the liver.

The dorsal aorta is a conspicuous vessel running along the middle line on the dorsal side of the body cavity. On its course backwards it gives off a large coeliaco-mesenteric artery which arises at the level of the pylorus and supplies the stomach, the liver, the intestine, and the spleen. Behind the coeliaco-mesenteric artery there arise close together a number of smaller arteries supplying the coiled intestine and the gonads. Farther back dorsal to the rectum a series of about six separate arteries supplying the rectum leave the dorsal aorta. These vessels are best seen with the viscera turned over to the animal's right side.

Urinogenital system

The position of the kidneys and the gonads has already been indicated.

In both sexes observe the adrenal bodies which consist of small patches of yellow tissue on the inner ventral surface of the kidneys on either side. In the region of the kidney the adrenal bodies are composed of both cortical and medullary tissue as in the higher vertebrates. Small bodies in front of the kidneys are also found which consist entirely of medullary tissue as in the supra-renals of the skate.

Male
(Fig. 16)

The testis is usually divided into two lobes. Stretch the mesentery connecting the testis with the kidney and the dorsal body wall. In this mesentery will be seen the vasa efferentia. These lead into a longitudinal duct lying along the anterior part of the kidney, and this longitudinal duct is connected with the anterior part of the kidney by numerous fine tubes.

The kidney is divisible into two regions, the narrow anterior region (sexual) and the wider oval posterior portion (excretory). Running along the outer border of the kidney is a conspicuous black or grey duct, the Wolffian duct. It extends forwards as a narrow thread past the anterior extremity of the kidney to the front end of the body cavity. This extension of the duct forwards beyond the kidney represents the portion of the duct connecting with the pronephros in the larva or tadpole. Stretch the mesentery connecting the Wolffian duct with the sexual portion of the kidney and observe the fine ducts leading from the kidney to the Wolffian duct (vasa deferentia). The posterior (excretory) part of the kidney is drained by larger ducts. Anteriorly one or two of these pass directly into the Wolffian duct. The remaining ones pass backwards parallel with the Wolffian duct and only open into it at its posterior end just before the Wolffian duct enters the cloaca. Coiled cloacal glands which form mucoid spermatophores lie beside the cloaca.

Female
(Fig. 17)

The ovary is paired and occupies the same relative position as the testis. It varies considerably in size; at its maximum the ovary extends over two-thirds of the body cavity length.

The paired oviduct is a long, much coiled duct situated on either side of the body cavity suspended by a mesentery from the middle line. It extends forwards to the extreme anterior end of the body cavity, where its internal opening may be seen. This opening is just beside the anterior end of the lung.

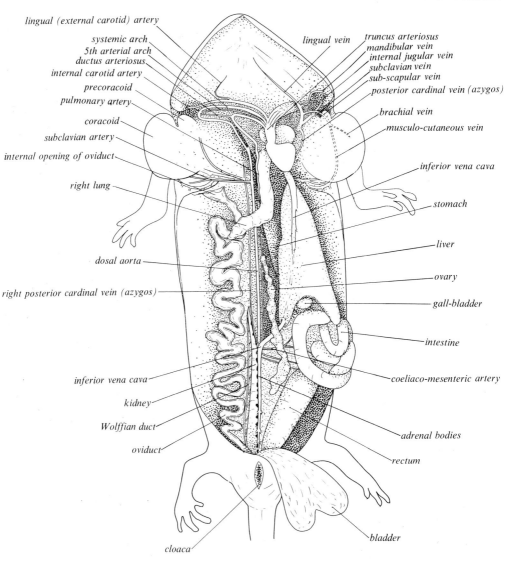

lingual (external carotid) artery

systemic arch
5th arterial arch
ductus arteriosus
internal carotid artery
precoracoid
pulmonary artery
coracoid
subclavian artery
internal opening of oviduct

right lung

dosal aorta

right posterior cardinal vein (azygos)

inferior vena cava
kidney
Wolffian duct
oviduct

cloaca

lingual vein

truncus arteriosus
mandibular vein
internal jugular vein
subclavian vein
sub-scapular vein
posterior cardinal vein (azygos)

brachial vein
musculo-cutaneous vein

inferior vena cava

stomach

liver

ovary

gall-bladder

intestine

coeliaco-mesenteric artery

adrenal bodies
rectum

bladder

FIG. 17. Dissection of a female salamander to show the contents of the body cavity; the viscera are turned over to the left to show the course of the posterior vena cava and the arterial and venous system round the heart.

Posteriorly the oviduct opens into the cloaca on either side of the rectum.

The kidney is the same shape as in the male and is similarly divisible into two regions; the narrow anterior region is not used for the passage of gametes, so that the whole kidney is solely excretory. Running along the outer border of the kidney is a narrow white duct. It is not coiled and it receives connecting ducts at intervals from the kidney. This duct is the Wolffian duct and corresponds with the black or grey duct of the same name seen in the male. The pronephric portion is not so easy to trace forwards as in the male. Posteriorly the Wolffian duct will be found to open into the cloaca just beside the opening of the oviduct.

The salamander is ovoviviparous and fertilization of the ova occurs within the oviduct, unlike the frog and newt in which the ova are externally fertilized and laid in water. In the spotted salamander (*Salamandra maculosa*) about twenty eggs undergo development in the posterior parts of each oviduct. In some female salamanders embryos may be found inside the oviducts which are distended in the lower region. The eggs are more heavily yolked than in the frog. Development proceeds until a young salamander is formed with four limbs and three pairs of external gills. The young lie in the egg membrane with the tail curled round the head. In this stage they are born into water where they live until metamorphosis occurs and the external gills are lost. Before birth the older embryos show an accumulation of yolk lying in the abdominal region.

In *Salamandra atra*, the black mountain salamander, internal fertilization also takes place, but only two embryos complete their development, one lying in the posterior part of each oviduct. In older stages the external gills are very long and are said to absorb a nutritive fluid excreted into the oviduct. At birth the external gills are shed and the young salamander of this species is entirely independent of the presence of water for its early growth.

Vascular system

The arteries and veins of the body cavity have already been seen.

The arteries and veins in the tail and posterior region of the body cavity

On one side of the cloaca carefully cut away the musculature in order to expose the posterior end of the kidney which lies dorsal to the cloaca. Find the caudal vein in the mid-ventral line. It bifurcates at the hind end of the kidney forming the renal portal veins. Follow one portal vein round the outer border of the kidney. At the level of the hind limb the portal vein receives two sciatic veins lying close together and just in front of them the femoral vein. All these vessels break up into capillaries, supplying the kidney.

Arterial blood is supplied to the kidney direct from the dorsal aorta.

From the femoral vein at the point where it enters the body

cavity arises the pelvic vein, which runs round the anterior border of the pelvic girdle. The two pelvic veins unite in the mid-ventral line to form the anterior abdominal vein.

Notice the dorsal aorta lying in the middle line between the renal portal veins and passing posteriorly dorsal to the caudal vein as the caudal artery. The dorsal aorta gives numerous renal arteries to the kidneys, and a small pair of iliac arteries leave the dorsal aorta above the kidneys and pass out to the hind limbs. The iliac arteries soon become embedded in muscle.

Arteries and veins in the region of the heart (Fig. 16)

In order to see the origin of the arteries and veins from the heart the pericardial membrane must be removed from the heart. This membrane is very tough and rather opaque. It lies closely pressed against the ventricle and must be removed with great care.

Note the sinus venosus which is formed by the union of the three venae cavae, two superior and one inferior. The inferior vena cava emerging from the anterior border of the liver has already been pointed out and its course backwards traced.

Follow one of the posterior cardinal veins forwards from the body cavity to the level of the pectoral girdle where the subclavian artery from the dorsal aorta crosses over it. Just at this point the posterior cardinal vein passes outwards and forwards to join with the superior vena cava. Just in front of the junction of the posterior cardinal with the vena cava, the superior vena cava splits into two.

The posterior branch runs dorsally and divides to form the internal jugular and the sub-scapular veins. The anterior branch divides again into two branches, the subclavian and the external jugular veins. The subclavian is the hinder of the two and passes directly outwards ventrally to the internal jugular. The subclavian is formed by the union of the brachial and musculo-cutaneous veins. The external jugular is formed by the union of lingual vein from the floor of the mouth and the external mandibular from the margin of the lower jaw. Find the musculo-cutaneous vein lying in the skin behind the posterior border of the coracoid. The paired subclavian arteries arise from the dorsal aorta opposite the anterior end of the liver and pass close to the brachial plexus.

The pulmonary veins lie along the inner border of the lungs. They unite and pass into the right auricle.

The heart and arterial arches (Fig. 17)

Note that the heart is essentially the same in appearance as that of the frog. There is a thick-walled ventricle behind and two thin-walled auricles in front. From the ventricle at its anterior end there arises the truncus arteriosus.

Let the blood out of the venous system by puncturing the auricles and the posterior vena cava. Squeeze out as much blood as possible, then wash under the tap and replace with fresh water. Now pass a roll of coiled paper down the throat in order to stretch the arterial arches.

Follow the truncus arteriosus forwards. Just before it leaves the pericardium it divides into four branches on either side. (The first two of the six branchial arteries of the embryo are lost in the adult.)

At their origin these branches are close together and are held by the tough pericardial membrane. In order to separate the arteries from each other it will be necessary to remove the pericardial membrane with a sharp, fine-pointed scalpel. Follow the arches outwards, removing the connective tissue and veins overlying them. The most anterior arch is the carotid. It passes horizontally outwards to the angle of the jaw. Half way along its course it gives off a small branch which runs forward (external carotid artery)[1] to supply the buccal cavity. Just beyond the origin of the external carotid is a thickening, the carotid gland. The arch continues as the internal carotid artery to the cranium.

The next arch, the systemic (fourth branchial of the embryo), runs parallel with the carotid arch, past the carotid gland and alongside the internal carotid artery. The further course of this arch is best traced by picking up the dorsal aorta and following this forwards to the point where it forks into the two systemic arches. This fork lies much farther forwards than in the frog. It will be found dorsal to the anterior end of the truncus arteriosus and dorsal to the oesophagus. Trace the connection between the systemic thus found with the systemic arch beside the carotid artery. Find the subclavian artery arising from the dorsal aorta just behind the union of the two systemic arches.

The next arch (fifth branchial) is often devoid of blood and is smaller than the systemic. If it is devoid of blood it can usually be found as a line of pigment lying in the connective tissue. Follow this arch outwards to its junction with the systemic arch. (In the newt the fifth arch in the adult is represented by a line of pigment only.)

The next arch is the pulmonary (sixth branchial). It lies parallel with the preceding arch, and about half way along this arch it turns backwards and runs along the outer border of the lung. At the point where the pulmonary arch turns backwards

[1] The internal carotids and external carotids (lingual arteries) are the roots of the original paired dorsal and ventral aortae respectively of the embryo. True external carotids are not found in modern fish, the ventral aorta ending blindly after giving off the anterior afferent branchial arteries.

a small vessel (ductus arteriosus) joins it with the systemic. The ductus arteriosus runs parallel with the preceding arch and is, in origin, a portion of the sixth branchial arch. The pulmonary arch does not give off a pulmocutaneous artery as in the frog, but several small arteries supplying the skin arise from the ductus arteriosus.

The pharynx and lungs

The position of the lungs in the body cavity has already been indicated. Cut open one lung with a pair of scissors and note the alveoli on the internal walls. Remove the heart, when the two bronchi will be seen through the dorsal wall of the pericardium as two red streaks. They unite at the level of the anterior border of the pericardium to form the trachea. The trachea is very short and opens into the mouth by the glottis behind which there is a small expansion which forms the larynx. The walls of the larynx are stiffened by the arytenoid cartilages. These cartilages may represent reduced gill arches. Pass a probe forwards from the lung up the bronchus and into the mouth in order to discover the glottis.

The hyoid arch is much reduced and the basal portion of it forms the hyoid cartilage, lying in the floor of the mouth. Cut through the angles of the jaw and open the mouth widely. Notice the absence of Eustachian tubes which is correlated with the absence of a tympanic membrane.

NOTE. The dissection of the nervous system is omitted. A dissection of the nervous system of the frog is given below.

5 *The muscles of the frog*

The use of electrical stimulators

THE actions of individual muscles can be observed by stimulating them electrically, using a frog which has been killed by 'pithing'. A simple stimulator can be made with two mounted needle electrodes, each wired to a terminal of a small rheostat which is connected to a 12-volt circuit transformed down from the ordinary laboratory electric supply. The technique for using this stimulator consists of placing the frog in a dissecting dish, exposing a muscle, inserting one needle electrode into some part of the frog's body, and touching the muscle with the other needle. The degree of contraction of the muscle depends on the strength of the current, which can be adjusted by using the rheostat or by altering the distance between the two needles. Ideally, the current should be adjusted so that the stimulated muscle contracts slowly but firmly while the surrounding muscles are not affected. This adjustment is sometimes difficult to achieve, for instance, with some of the leg muscles. If the muscles begin to dry up and stop reacting, they can be revived by moistening them with 'frog ringer' solution; this should be done cautiously since excess of this solution will cause the stimulus to spread over several muscles. Generally, the frog's muscles should continue to respond to stimulation for at least 2 hours after 'pithing'.

It is possible, of course, to dissect the muscular system of the frog without using electrical stimulation, but the actions of the individual muscles are more obvious and easier to determine when they contract under stimulation. If a fresh frog is used without a stimulator, it should be soaked in water so that the muscles swell and become more distinct. If no fresh frogs are available, specimens preserved in formalin are recommended.

The muscles of the head
(Figs. 18, 19)

Remove the skin from the dorsal surface and one side of the head and prop the frog up in an erect position. Open the

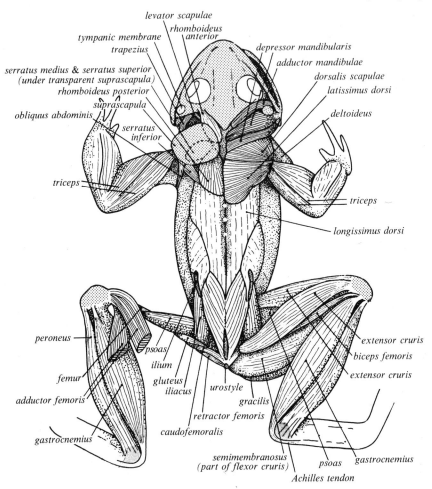

FIG. 18. Frog: dorsal view showing the superficial muscles on the right and the deeper muscles on the left. The skin and some connective tissue have been removed. Exposed or nearly exposed bone and cartilage are shown by mechanical stippling. On the left side the following muscles have been removed: depressor mandibularis, adductor mandibulae, dorsalis scapulae, latissimus dorsi, deltoidus, extensor cruris, biceps femoris, flexor cruris, and part of adductor femoris.

mouth with a scalpel, free the lips of mucus, and allow the mouth to close again. Place one electrode in the viscera and with the other touch the *depressor mandibularis* which arises from the neck and skull and passes posterior to the tympanic membrane to insert on the mandible. This muscle opens the mouth. Anterior to the tympanic membrane is the *adductor mandibulae* which closes the mouth. Touch each of the muscles

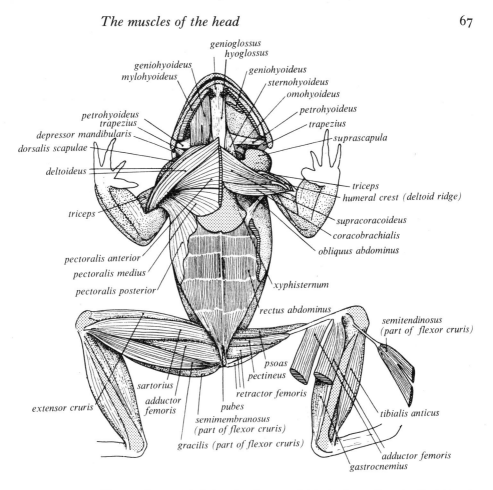

FIG. 19. Frog; ventral view showing superficial muscles on the animal's right and deeper muscles on the left. The skin and some connective tissue have been removed. Exposed or nearly exposed bone and cartilage are shown by mechanical stippling. The mylohyoideus is cut away on both sides, and on the left the following muscles have been removed either completely or in part: geniohyoideus, depressor mandibularis, deltoideus, dorsalis scapulae, pectorales anterior, medius, and posterior, extensor cruris, sartorius, gracilis, and semimembranosus. The adductor femoris and semitendinosus have been reflected.

in turn with the electrode so that the mouth is made to open and close alternately. Cut through the adductor mandibulae muscles on both sides of the head and remove the depressor mandibularis on one side. This will expose the *trapezius*, arising from the otic region and inserted on the anterior border of the scapula. When stimulated, this muscle either bends the neck and rotates the head or protracts the pectoral girdle.

Place the frog on its back and carefully remove the skin from the ventral surface of the throat to expose the *mylohyoideus*, a thin sheet of muscle between the two mandibles. Place one electrode in the viscera, touch this muscle with the other and observe that it contracts, raising the floor of the mouth. Cut it away, taking care not to injure the transparent muscles lying dorsal to it. The two *geniohyoideus* muscles are now exposed. Each of the muscles, which are divided into several slips, arises near the midline from the mandible and forks posteriorly to insert on the posterior end of the hyoid plate. A *sternohyoideus* muscle is inserted between the forked ends of each geniohyoideus and arises from the dorsal surface of the coracoid and sternum. Stimulate these muscles and observe that the geniohyoideus protracts the hyoid plate and the sternohyoideus retracts it. Lateral to the sternohyoideus is the *omohyoideus*, arising from the scapula and inserted on the lateral edge of the hyoid plate which it draws backwards and laterally. Put one electrode on the sternohyoideus and the other on the geniohyoideus of the same side and observe that when both contract together they open the mouth. Cut away the geniohyoideus, sternohyoideus, and omohyoideus muscles.

The *petrohyoideus* muscles consist of four small slips on each side of the head arising from the otic region anterior to the trapezius and inserted on the lateral edge of the hyoid plate. When they contract they raise the hyoid plate and compress the throat.

Movements of the floor of the mouth and of the hyoid plate play an important part in both the buccopharyngeal and the pulmonary respiration of the frog. If the nostrils remain open, air is drawn into the mouth by downward movement of the hyoid plate and is expired by upward movement of the hyoid. When the nostrils are closed and the glottis is open, air can be forced from the mouth into the lungs by upward movement of the hyoid and can be drawn from the lungs into the mouth by its downward movement. When the frog is in an upright position, the hyoid plate (and with it the floor of the mouth) is lowered by the sternohyoideus and omohyoideus and raised by the petrohyoideus assisted by the geniohyoideus muscles.

The frog feeds by rapidly protracting the tongue and then retracting it with an insect or woodlouse adhering to its sticky tip. The protractor of the tongue is the *genioglossus* and the *hyoglossus* is the retractor; it is difficult to stimulate these muscles separately. The genioglossus arises from the median symphysis of the mandibles and passes dorsally into the tongue while the hyoglossus occupies the ventral surface of the hyoid plate, arising from its posterior cornua and passing dorsally anterior to the hyoid plate into the tongue. Prop open the mouth and note the resting position of the tongue. Stimulate each of the two muscles by using both electrodes close together; it should be possible to make the tongue turn forwards and protract a short distance and then retract and turn backwards.

The muscles of the limbs and girdles
(Figs. 18, 19)

In this and the following chapters, the actions of the limb muscles will be described in the following terms, which assume that the vertebral column is horizontal: a *levator* pulls a bone in the dorsal direction and a *depressor* moves it ventrally, a *protractor* moves the bone in the direction of the head and a *retractor* moves it posteriorly, a *rotator* turns the bone about its long axis, a *flexor* decreases the acute angle between two long bones and an *extensor* increases this angle. Many muscles have composite actions but they can be described by combining these terms, e.g. a depressor–protractor pulls a bone ventrally and towards the head.

The muscles of the pectoral girdle and fore limb

Remove the skin from the dorsal and ventral surfaces of the body in the region of the pectoral girdle and from one fore limb. Lay the frog on its back and stick one electrode through one of its feet. Note the *rectus abdominis* muscle in the mid-ventral line arising from the pubic symphysis and inserted on the dorsal surface of the sternum. From the lateral surface of the rectus arises the posterior *pectoralis* which inserts on the humeral crest. Stimulate the posterior pectoralis with the other electrode and observe that it retracts, depresses, and rotates the humerus. The anterior and median parts of the pectoralis arise from the sternum and are also inserted on the humeral crest. When these parts are stimulated, they depress and rotate the humerus. The *deltoideus* lies in front of the anterior pectoralis and arises from the anterior edge of the coracoid and from the episternum and inserts on the humeral crest. Stimulate the deltoideus and observe that it is a protractor and rotator of the humerus.

Prop the frog up on its side and observe the *latissimus dorsi*, arising from the vertebral spines and inserted on the humeral crest. Stimulate this and observe that the humerus is retracted, elevated, and rotated. Remove the depressor mandibularis to expose the *dorsalis scapulae*, which arises on the dorsal surface of the suprascapula and is a levator and rotator of the humerus. Observe the antagonistic arrangement of the following muscles: the dorsalis scapulae against the anterior and median pectoralis; the deltoideus against the latissimus dorsi and posterior pectoralis. All these muscles are inserted on the humeral crest and rotate the humerus about its long axis when they contract.

Carefully remove the anterior, median, and posterior pectoralis muscles to expose the *supracoracoideus* and *coracobrachialis*. The latter arises from the posterior edge of the coracoid and inserts on the humerus, which it retracts. The supracoracoideus arises on the ventral surface of the coracoid and converges to a strong, glistening tendon which passes along a groove beside the humeral crest and inserts on the radius. Stimulate the muscle and observe that it flexes the elbow joint. This joint is extended by the *triceps* which has one head on the scapula and three heads on the humerus.

The muscles of the forearm can now be investigated with the stimulator but they will not be described in this manual.

The muscles described so far have been concerned with movements within the fore limb and of the fore limb relative to the pectoral girdle and body. The pectoral girdle is itself movable in relation to the rest of the body and the muscles that move it can now be investigated.

The *sternohyoideus, omohyoideus*, and *trapezius* have already been described as muscles of the head but they can also protract the pectoral girdle. The *rectus abdominis*, already mentioned, may retract the girdle.

Remove the latissimus dorsi and the dorsalis scapulae. The anterior and posterior *rhomboideus* muscles protract and retract the suprascapula, respectively, and are inserted on its ventral surface near its upper edge. The former arises from the posterior edge of the skull and the latter from the neural spines of the third and fourth vertebrae and from connective tissue over the *longissimus dorsi*, the mass of muscle that lies immediately dorsal to the vertebral column along the length of the body. Raise the suprascapula and observe the *levator scapulae* arising on the exoccipital and inserting on the ventral surface of the suprascapula in two parts. On stimulation, these muscles elevate and protract the scapula. The three *serratus* muscles arise from the transverse processes of the third and fourth vertebrae and have the following insertions: serratus superior beside rhomboideus posterior; serratus medius in the middle of the ventral surface of the suprascapula; and serratus inferior on the posterior border of the scapula. The first two are retractors of the scapula and serratus inferior is mainly an elevator. A part of the *obliquus externus* muscle of the body is inserted on the posterior border of the scapula, which it retracts and depresses.

The muscles of the pelvic girdle and hind limb

The frog normally moves by leaping; the propulsive movement is the result of simultaneous retraction of both femora and extension of both knee joints. The hind limb is a highly specialized structure and the muscles are complicated. Many of them pass over two joints. The pelvic girdle is fixed to the sacral vertebra and there are no muscles to move it relative to the body.

Remove the skin from one hind limb and from over the two ilia. Open the abdomen and cut the nerves of the sciatic plexus before using a stimulator. Lay the frog on its back with the leg extended and gently stimulate the *sartorius*, a thin slip of muscle arising from the pubic symphysis and inserted on the tibia. Observe that the femur is protracted and depressed and the knee is flexed. Turn the frog over and find the *biceps femoris*, arising from the ilium and inserted on the tibiofibula. Stimulate this muscle and note that the femur is protracted and elevated and the knee is flexed. Cut away the sartorius and the biceps femoris.

Investigate the *extensor cruris*, which extends the knee joint and can protract the femur. This muscle has three parts, two arising from the ilium and the third from the edge of the acetabulum; these three parts join to form a powerful muscle which lies on the anterior surface of the femur and becomes tendinous and stretches over the knee joint to insert on the tibia. Stimulate the muscle and then cut it away. The *flexor cruris* group of three muscles are the principal retractors of the femur. Semi-membranosus and gracilis both arise from the posterodorsal edge of the ischium and insert below the knee, the former being dorsal to the latter. Stimulate them and cut them away to expose semitendinosus, which arises by two short tendons from the ischium and forms a fleshy muscle inserted by a short tendon on to the tibia. Stimulate it and observe that it retracts and partly depresses the femur and may flex the knee. Cut it away.

Stimulate the *adductor femoris*, which has three parts (two of which are difficult to separate) arising from the ischium and pubis and inserted on to the distal end of the femur. Observe that the femur is depressed and then cut the muscles away. Find the short *retractor femoris* which also consists of three parts, all arising from the ischium and inserted on to the proximal end of the femur. Stimulate this and then cut this away. The *pectineus* is a short depressor of the femur and arises from the pubis while *psoas* is a short protractor arising from the postero-ventral border of the ilium. Two small muscles, *gluteus* and *caudofemoralis*, arise on the posterior ends of the ilium and urostyle, respectively, and insert on the femur of which they are weak levators. Remove all these muscles.

In contrast to the fore limb, where all the long muscles from the girdle to the humerus rotate the limb about its long axis, rotation of the femur is produced only by two small muscles, the *iliacus* and the *obturator internus*. The former arises from the dorsal border of the ilium and forms a strong tendon inserted on the posterior face of the femur, while the latter has a long origin round the acetabulum and inserts on the dorsal surface of the femur. They are difficult muscles to stimulate.

Observe and stimulate the extensor and flexors of the ankle-joint. The former is the *gastrocnemius* with the Achilles tendon over the heel, and the latter are *peroneus* and *tibialis anticus*.

A LIST OF THE NAMES FOR MUSCLES USED IN THIS MANUAL WHICH DIFFER FROM THOSE USED BY GAUPP[1] (1896)

PRESENT NAME	GAUPP'S NAME
Adductor mandibulae	Masseter major and minor, pterygoideus, and temporalis
Biceps femoris	Ileofibularis
Caudofemoralis	Pyriformis
Extensor cruris	Triceps femoris
Flexor cruris	Gracilis major and minor, semimembranosus, and semitendinosus
Gastrocnemius	Plantaris longus
Gluteus	Ileofemoralis
Iliacus	Iliacus externus
Mylohyoideus	Submaxillaris and subhyoideus
Psoas	Iliacus internus
Retractor femoris	Obturator externus, quadratus femoris, and gemellus
Supracoracoideus	Coracoradialis
Trapezius	Cucullaris
Triceps	Anconaeus

[1] Gaupp, E. (1896), A. Ecker's and R. Wiedersheim's *Anatomie des Frosches*, Abt. i.

6 *The central and sympathetic nervous system of the frog*

General directions

OPEN a fresh frog in the usual manner by pinning it out in a dish and cutting through the body wall in the midventral line. Cut through the pectoral girdle and pin back the two halves of the girdle on either side. Distend the oesophagus by passing a roll of paper down the throat. Cut open the heart and let the blood drain out.

In the case of a female carefully remove the ovaries and oviducts. Do not remove the alimentary canal, but the liver can be cut away if desired. After these preliminaries refill the dish with clean water. The nerves can be made more obvious by pouring 70 per cent alcohol on to them before filling the dish with water but this procedure has the disadvantage of making the nerves brittle.

Glossopharyngeal nerve
(Fig. 20)

Opposite the angle of the jaw may be seen two small white nerves passing forwards below the mylohyoid muscle between the rami of the lower jaws. These nerves supply the tongue region. The posterior of the two is the hypoglossal or first spinal nerve. The anterior is the IX (glossopharyngeal) nerve. Below the mylohyoid muscle the hypoglossal nerve lies superficial to the glossopharyngeal nerve. Follow these two nerves forwards to their destination and backwards to their origin from the central nervous system. As they pass over the arterial arches running out from the pericardium they turn dorsally and then run forwards looping upon themselves. Follow the glossopharyngeal nerve forwards and dorsally towards the angle of the jaw, gently loosening the connective tissue in this region. This nerve will be found lying close to the internal carotid artery and emerging from the back of the skull.

The vagus nerve
(Fig. 20)

The internal jugular vein lies close to the internal carotid artery. Gently stretch this region and the greyish X nerve (vagus) will be seen lying close to the internal jugular vein and glossopharyngeal nerve. The vagus leaves the posterior part of the skull along with the glossopharyngeal nerve. Carefully follow the vagus nerve forwards to its origin from the skull. Just after its origin a ganglionic enlargement will be seen on the nerve.

The various branches of the vagus nerve must now be traced. In doing this care must be taken not to injure any part of the hypoglossal nerve which has already been noticed. The vagus nerve passes backwards lying dorsal to the cutaneous artery.

Close to its origin from the skull the vagus gives off the recurrent laryngeal nerve. This is a slender nerve which lies at first parallel with the vagus near its ventral side. It then passes round the posterior cornu of the hyoid and over the ventral side of the pulmocutaneous artery. It then loops round this artery close to its origin from the conus arteriosus and passes dorsal to the arterial arches to reach the larynx in the middle line.

Just before the main branch of the vagus passes ventral to the systemic arch it gives off gastric branches, usually two in number. These lie along the sides of the oesophagus and pass backwards to the stomach.

The vagus then passes between the cutaneous artery and the systemic arch, running dorsal to the pulmonary artery and passing close to the anterior border of the lung. One or two pulmonary branches leave the vagus close to the pulmonary artery and run along this artery to supply the lung.

The vagus nerve passes on as the cardiac branch running into the pericardium to supply the walls of the sinus venosus and the interauricular septum.

Spinal nerves[1]
(Fig. 20)

The hypoglossal nerve (first spinal) has already been seen in the dissection of the glossopharyngeal nerve. It can now be traced to its origin from the spinal cord emerging between the first and second vertebrae just in front of the subclavian artery.

The remaining spinal nerves may now be traced. In order to do this lay the viscera over to one side. The kidneys lie attached to the dorsal body wall within a large lymph space behind the peritoneum. Carefully cut through the peritoneum forming the lateral wall of this lymph space between the lateral border of one kidney and the body wall. Carefully turn the outer border of the kidney over towards the viscera, so uncovering the roots of the spinal nerves and the dorsal aorta.

There are ten spinal nerves and the first (hypoglossal) has already been seen. The second and third spinal nerves leave

[1] Some authorities consider that the true first spinal nerve is absent and therefore number these nerves II to XI.

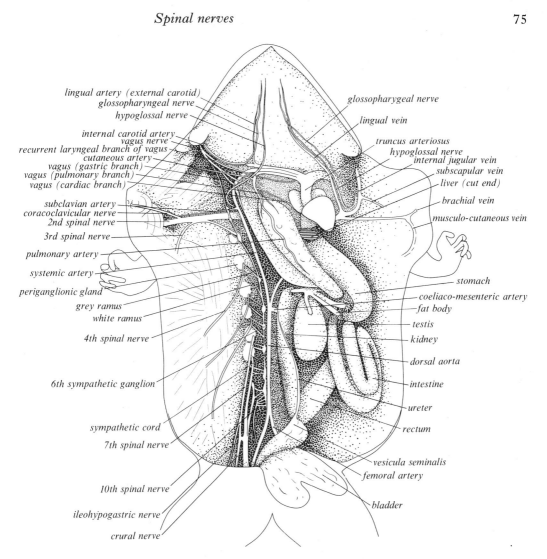

lingual artery (external carotid)
glossopharyngeal nerve
hypoglossal nerve
internal carotid artery
vagus nerve
recurrent laryngeal branch of vagus
cutaneous artery
vagus (gastric branch)
vagus (pulmonary branch)
vagus (cardiac branch)
subclavian artery
coracoclavicular nerve
2nd spinal nerve
3rd spinal nerve
pulmonary artery
systemic artery
periganglionic gland
grey ramus
white ramus
4th spinal nerve
6th sympathetic ganglion
sympathetic cord
7th spinal nerve
10th spinal nerve
ileohypogastric nerve
crural nerve

glossopharygeal nerve
lingual vein
truncus arteriosus
hypoglossal nerve
internal jugular vein
subscapular vein
liver (cut end)
brachial vein
musculo-cutaneous vein
stomach
coeliaco-mesenteric artery
fat body
testis
kidney
dorsal aorta
intestine
ureter
rectum
vesicula seminalis
femoral artery
bladder

FIG. 20. Frog dissected to show the sympathetic system with the ninth and tenth cranial and the ten spinal nerves. The throat has been distended with a plug, and the liver removed.

the spinal cord between the second and third, and third and fourth vertebrae respectively.

The second spinal nerve is very large. The brachial plexus is formed by the fusion of one or two small branches from the base of the hypoglossal nerve, the second spinal nerve, and a branch from the third spinal nerve. These branches unite to form the brachial nerve. The brachial nerve gives off a large branch, the coracoclavicular, about ⅓ inch from its origin. This branch supplies the shoulder muscles. The brachial nerve then runs down the arm, dividing just above the elbow into the radial and ulnar nerves supplying the forearm and hand.

The third spinal nerve runs closely associated with the brachial plexus for a short distance. After giving off a branch to this plexus the third spinal nerve separates from it and passes backwards and outwards to supply the musculature and skin of the body wall.

The fourth, fifth, and sixth spinal nerves are small and supply the musculature and skin of the body wall. They leave the vertebral canal between the fourth and fifth, fifth and sixth, and sixth and seventh vertebrae respectively. Trace these nerves outwards to their destination.

The seventh, eighth, and ninth spinal nerves are much larger and together form the sciatic plexus. These nerves leave the vertebral canal between the seventh and eighth, eighth and ninth, and ninth vertebra and urostyle respectively. From their origin from the vertebral column they run posteriorly and after about ¾ inch of their course unite to form the sciatic plexus opposite the middle of the urostyle. From the sciatic plexus branches are given off to the rectum, bladder, and oviducts and surrounding organs. Just before the seventh nerve joins the plexus it gives off the small ileohypogastric nerve and the larger crural nerve supplying the muscles and skin of the abdomen and thigh. Beyond the plexus is a large sciatic nerve which runs down the thigh. Follow this nerve backwards from the body cavity into the leg as far as possible. Then either cut away the femur in order to discover the further course of the nerve or find the nerve again on the inner side of the leg by carefully removing the muscles. The sciatic nerve gives branches to the muscles of the limb and a short distance above the knee it divides into the tibial and peroneal nerves supplying the leg and foot and the calf muscle (gastrocnemius muscle).

There is considerable variation in the size of the nerves and the method of their union in forming the sciatic plexus.

The tenth spinal or coccygeal nerve is very much smaller than the others, it emerges through a small foramen in the urostyle near its anterior end. This nerve then passes directly backwards close to the middle line and gives branches to the cloaca, bladder, and surrounding organs, and also contributes a small branch to the sciatic plexus.

Round the roots of the spinal nerves just as they emerge from the vertebral canal will be seen white calcareous patches which are formed by the periganglionic glands or 'glands of Swammerdam'.

In order to see the roots of the spinal nerves these glands must be removed. This dissection should be postponed until after the examination of the sympathetic nervous system.

The sympathetic nervous system (Figs. 20, 21)

This system consists of a longitudinal strand of nervous tissue lying on either side of the body and connected by branches with the central nervous system. The two main sympathetic trunks may be found lying on either side of the dorsal aorta. The course of these trunks is often rendered conspicuous by the large number of pigment cells that lie over them. The sympathetic trunk lying close to the posterior part of the systemic arch and the dorsal aorta will be easily seen. Follow the trunk forwards. Ganglionic enlargements are present along it. These ganglia are segmental in arrangement. The first two trunk ganglia lie very close together at the base of the first and second spinal nerves, and the ninth and tenth ganglia are practically fused. Gently pull the kidney and dorsal aorta sideways in order to stretch the connections of the sympathetic chain.

These connections, the rami communicantes, are small nerves passing from the spinal cord to the sympathetic trunk and from the sympathetic trunk to the spinal nerves.

The 'white rami' leave the spinal cord through the intervertebral notches and pass to the sympathetic trunk usually uniting with it just in front or behind the ganglia. These white rami are composed of medullated fibres, but usually appear very dark in consequence of the abundant covering of pigment cells.

The 'grey rami' leave the sympathetic ganglia and pass outwards to unite with the spinal nerves a short distance from their origin from the vertebral canal. They are composed of non-medullated sympathetic fibres which pass out with the spinal nerves to the peripheral parts of the body. In the posterior region the grey rami of each spinal nerve may be duplicated. The small tenth spinal nerve may receive as many as twelve branches from the sympathetic system. Some rami may be absent or mixed.

Now follow the sympathetic chain forwards towards the head. On reaching the brachial plexus the sympathetic passes away from the systemic arch and runs forwards round the subclavian artery towards the origin of the vagus. The sympathetic here is very small and difficult to trace. It runs forwards and supplies the head (Fig. 21a).

From the sympathetic ganglia nerves are given off to the viscera and blood-vessels. The cardiac plexus is formed by nerves arising from the first sympathetic ganglion near the base of the hypoglossal nerve. These sympathetic nerves are

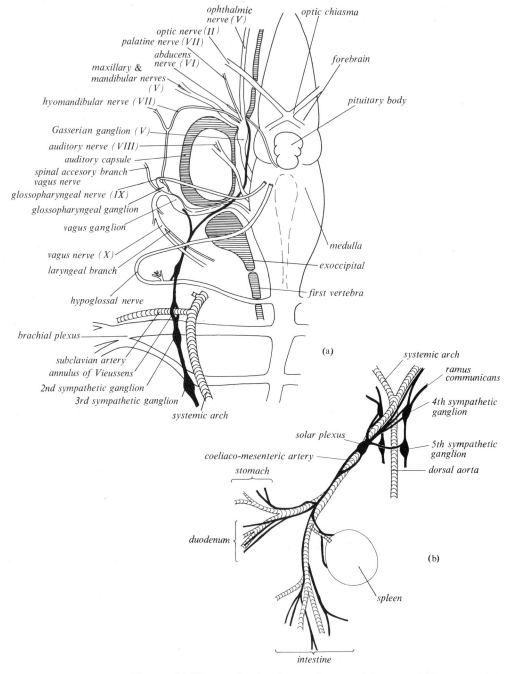

FIG. 21. (a) Diagram showing the anterior part of the sympathetic system of the frog with its connections to the cranial nerves from the ventral side.

(b) Diagram showing the origin of the solar plexus from the sympathetic trunks of the frog.

In both diagrams the cut bone is cross-hatched, the blood-vessels striped, and the sympathetic nerves black.

very small and difficult to trace. They form a meshwork over the auricles and round the openings of the arteries and veins into the heart.

The solar plexus lies around the coeliaco-mesenteric artery (Fig. 21b). Its nerves are derived mainly from the third, fourth, and fifth sympathetic ganglia which are associated with the third, fourth, and fifth spinal nerves.

The spinal cord

Remove the calcareous patches at the base of the spinal nerves and expose the roots of these nerves as fully as possible. Now turn the animal over and remove the skin and musculature from the mid-dorsal region. Carefully cut away the neural arches and slice away the dorsal part of the urostyle so as to expose the whole of the spinal cord. Gently draw it away from one side of the vertebral column and see the dorsal and ventral roots of each spinal nerve.

Behind the level of the exit of the seventh spinal nerve, the spinal cord narrows considerably as it gives off the eighth, ninth, and tenth nerves, after which it passes backwards as the filum terminale lying in the posterior part of the vertebral column and in the urostyle. The roots of the seventh, eighth, ninth, and tenth nerves lie beside the filum terminale within the vertebral column for a short distance, so forming the cauda equina. The exits of these nerves from the vertebral column have already been seen.

Notice the continuations of the periganglionic glands within the vertebral canal.

[It is assumed that the student will have dissected the brain and the remaining cranial nerves in a course of elementary biology.]

7 *The lizard*

Lacerta viridis

External features   NOTE that the general appearance of the body is very like that of the salamander. The limbs are a little longer but hardly raise the belly off the ground. The surface of the body is covered with horny scales arranged in a regular pattern. On the head the scales are large and closely applied to the roofing bones of the skull, but the position of these scales does not correspond with that of the underlying bones. The scales on the under surface are larger than the scales on the upper surface. The body behind the head region is slightly constricted to form the neck. The tail is long and tapering, being about twice the length of the head and trunk together. The mouth is terminal as in the salamander. The paired external nares lie on either side at the tip of the snout. The eyes lie on either side of the head and are provided with movable eyelids which can completely cover over the eye. A nictitating membrane is also present. Behind the angle of the jaw lies the tympanic membrane freely exposed on the surface. (In the skink the tympanic membrane is covered by a fold of skin bearing scales.) Open the mouth and observe the small teeth on the upper and lower jaws. On the floor of the mouth lies the bifid tongue which is attached posteriorly. On the fore and hind limbs there are five long digits each ending in strong claws. The fourth digit on the hind limb is much longer than the rest. Just behind the hind limb lies the cloaca which appears as a transverse slit. A large transverse scale, the pre-anal plate, lies in front of the cloaca.

General directions for dissection   Pin the animal down on its back by four pins through the limbs. Open the body cavity by a median ventral incision extending from the pectoral to the pelvic girdle. Notice the anterior abdominal vein lying in the body wall midventrally and passing forwards into the liver. Cut through this vein and lay back the body wall on either side, so exposing the viscera.

Now cut through the pectoral girdle carefully in the middle line, and carry the cut forwards to the chin. The pelvic girdle is very much wider antero-posteriorly than in the salamander. Just in front of the pelvic girdle are a pair of compact yellow fat bodies. These will be seen to lie between the peritoneum and the musculature, and are not suspended in the body cavity by mesentery as in Amphibia. Cut through the pelvic girdle in the middle line, leaving the black peritoneum from the ventral side of the body cavity uncut. Pin back the two halves of the girdle on either side. In doing this notice the pelvic veins running along the anterior border of the pelvic girdle and uniting to form the median abdominal vein. The pelvic veins lie embedded in the fat bodies and receive branches from them.

Contents of the body cavity
(Fig. 22)

The general arrangement of the viscera is very similar to that of the salamander.

The peritoneum lining the body cavity is more strongly pigmented than in the salamander. In the posterior part of the body cavity the peritoneum is completely black. In the anterior part there is little pigment and a sharp line of demarcation separates the two regions.

In the anterior part of the body cavity lies the liver. The inferior vena cava emerges through the liver at its anterior end and passes forwards towards the heart, lying freely exposed for about three-quarters of an inch. The gall-bladder can be seen lying in a furrow on the right lobe of the liver. On either side of the liver will be seen the paired lungs lying in the anterior part of the body cavity. Gently turn the liver over towards the animal's left side. The stomach will now be exposed lying dorsal to the liver. The stomach is a tubular organ slightly greater in diameter than the intestine and may often be found to be folded on itself. Between the stomach and the duodenum lies the pyloric constriction. The duodenum runs obliquely forwards towards the hind end of the liver.

In the mesentery between the stomach and the duodenum lies the pancreas. The pancreas is a long whitish gland and is continued forwards to the posterior face of the left lobe of the liver. At this point the bile-duct reaches the pancreas. Carefully remove the substance of the liver lying over the gall-bladder in this region and find the bile-duct leaving the gall-bladder and running through the pancreas. The bile-duct continues through the substance of the pancreas receiving the pancreatic ducts and enters the duodenum.

Notice the gastro-hepatic omentum, a thin sheet of peritoneum extending between the dorsal surface of the liver and

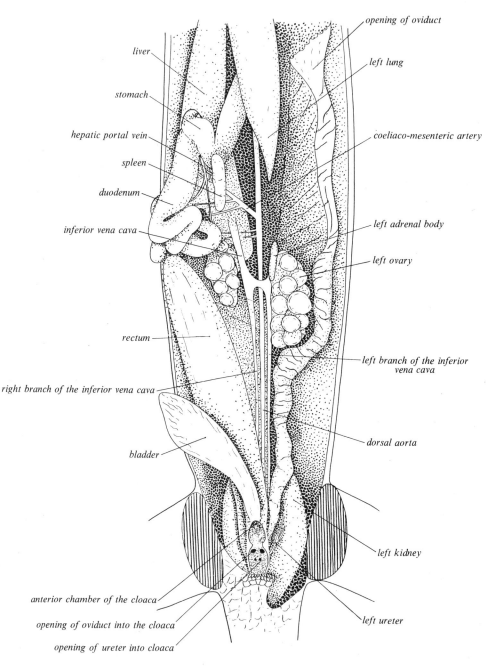

FIG. 22. Contents of the body cavity of a female lizard with the viscera turned over to the right and the cloaca cut open ventrally.

the ventral surface of the stomach. It is continued backwards as the duodenohepatic omentum which connects the liver with the first lobe of the intestine. The duodenum is continued as the small intestine which is much coiled and suspended by a mesentery. The small intestine widens into the rectum which may be folded on itself and is about 3 inches in length. The rectum is suspended from the body wall by a mesentery. Follow the rectum backwards to the cloaca.

Ventral to the rectum will be seen the bladder lying in the peritoneum which covers the ventral abdominal wall. Carefully move the bladder aside in order to follow out the rectum.

The spleen will be found lying in the mesentery connecting the stomach with the dorsal body wall. It is much smaller than in the salamander, about $\frac{1}{2}$ inch in length, and lies near the pylorus.

On either side of the rectum lie the paired gonads. Usually the right gonad lies slightly anterior to the left. In the female the oviduct lies on the outer side of the ovary and in the male the reduced mesonephros and its duct lies close to the testis.

The paired kidneys are much more compact than in the salamander, and are situated farther back in the body cavity opposite the bases of the hind limbs. These kidneys, unlike those of the salamander, are metanephric kidneys.

Arteries and veins of the body cavity

The hepatic portal vein may be found running through the pancreas. It enters the liver near the point where the bile-duct passes into the pancreas. The hepatic portal vein is formed by the union of the vessels draining the stomach, duodenum, small intestine, spleen, and rectum.

Turn the viscera over to the animal's left side. The inferior vena cava will be seen entering the right dorsal lobe of the liver just as in the salamander. Follow this vein backwards. It runs to the inner side of the right ovary or testis. Opposite the middle of the left ovary or testis it divides into two. Follow each of these vessels backwards. They will be seen to arise on the inner sides of the metanephric kidneys. Note that the posterior vena cava is not united with posterior cardinal veins as in the adult salamander.

The dorsal aorta may be seen lying along the body wall in the mid-dorsal line. The large coeliaco-mesenteric artery leaves the dorsal aorta at the level of the spleen and supplies the stomach, liver, spleen, and intestine. Behind the coeliaco-mesenteric artery a pair of arteries pass to the gonads, and a few small branches from the dorsal aorta supply the viscera and body wall.

Urinogenital system (Fig. 22)

In both sexes observe the compact brown adrenal bodies lying on the inner side of the gonads close to the branches of the inferior vena cava.

Male

The paired testes have already been noticed. They are round yellowish organs lying suspended in the body cavity, the left testis slightly behind the right.

Along the inner border of each testis lies the epididymis which is the reduced mesonephros and its Wolffian duct. The epididymis is mainly composed of the long convoluted Wolffian duct which now becomes the vas deferens and collects the spermatozoa from the testes. Follow the vas deferens backwards. Posteriorly it lies over the ventral surface of the kidney and runs backwards opening into the cloaca. Remove, if necessary, the body wall at the side of the kidney in order to expose it fully.

The kidney is a compact brown body lying opposite the base of the hind limbs. Posteriorly the kidney projects slightly behind the cloaca into a pouch from the body wall. Carefully pull the rectum to one side to expose fully the course of the vas deferens. Below the peritoneum covering the kidney may be seen the narrow ureter draining the kidney and passing into the cloaca just behind the vas deferens. It may be necessary carefully to remove the peritoneum covering the kidney in this region. This ureter drains the metanephric kidney only and is not associated with the genital organs.

On either side of the cloaca attached to the edges of its posterior border are a pair of penes. These are hollow sacs which are extrusible through the cloaca.

Slit open the cloaca in the mid-ventral line. Note that just behind the point where the bladder enters, it is divided by an incomplete transverse partition. Into the anterior compartment of the cloaca open the bladder and the rectum, while into the posterior portion open the ureters and the vasa deferentia.

Female

The ovaries are paired and lie suspended in the body cavity in the same relative position as the testes. The oviducts are delicate, pleated tubes lying closely attached to the wall of the body cavity. They extend forward anteriorly to the level of the middle of the lung. At the anterior end the walls of the tube are very transparent and the tube itself widens. Find the internal opening of the oviduct by inserting a probe. Follow the oviduct backwards and find its opening into the posterior chamber of the cloaca beside the ureter.

The kidney and the ureter are the same as in the male.

Vascular system

The arteries and veins of the body cavity have already been seen (p. 83).

*Arteries and veins
of the tail and
posterior region of
the body cavity*
(Fig. 22)

The veins and arteries in the posterior region of the body and
the tail resemble those of the salamander. The renal portal
system is however less well developed. This may be correlated
with the presence of adult metanephric kidneys. The caudal
vein bifurcates to form the renal portal veins which receive
sciatic and femoral veins from the limbs. The pelvic veins
have already been seen and arise from the junction of the renal
portal and sciatic veins. Renal veins pass to the kidneys from
the renal portal veins. The dorsal aorta gives off a pair of iliac
arteries to the hind limbs, and several renal arteries to the
kidneys. The dorsal aorta continues as the caudal artery.

*Arteries and veins
in the region of the
heart*
(Fig. 23)

Remove the pericardium from the heart, disclosing the ventricle behind and the two auricles on either side in front. In
preserved specimens the auricles are usually very brittle and
cannot be pushed aside without breaking them off. Trace the
inferior vena cava forwards to the junction with the two
superior venae cavae to form the sinus venosus. The sinus
venosus can be seen by gently raising the right auricle. The
sinus venosus discharges its blood into the right auricle.

Follow the superior venae cavae forwards. The left superior
vena cava is formed by the junction of the large internal jugular vein and the subclavian vein. The internal jugular is very
large and drains the contents of the cranium. It lies along the
lateral border of the neck.

Follow the right superior vena cava forwards. Almost immediately after leaving the sinus venosus it receives two
vessels, the anterior of which is the subclavian and drains the
fore limb, while the posterior is the azygos and drains the
intercostal muscles. From the junction with the vena cava the
azygos runs vertically upwards until it reaches the dorsal wall
of the body cavity. On reaching the body wall it turns backwards and runs posteriorly alongside the vertebral column
embedded in the musculature. It is unpaired and drains the
intercostal muscles on both sides. It represents the right
posterior cardinal sinus of lower forms. A short distance in
front of the subclavian vein the superior vena cava receives
the internal and external jugulars. The internal jugular vein
resembles in size and position that of the left side. Note the
external jugular lying alongside the trachea on the right side.
It joins the right superior vena cava, but its connection with
this vessel cannot be traced until after the arterial system has
been investigated. The external jugular is an unpaired vessel
and drains both the right and left sides.

Gently raise the heart and turn it over to the right-hand side.
Carefully dissect away the floor of the pericardium and observe the pulmonary vein draining both lungs and running

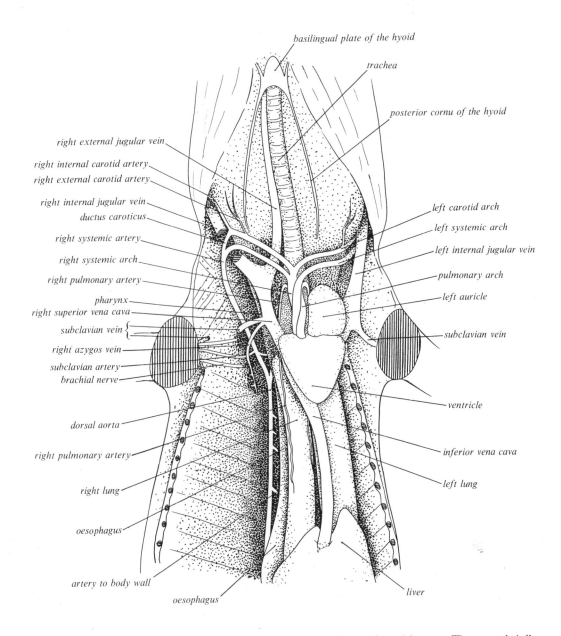

basilingual plate of the hyoid

trachea

posterior cornu of the hyoid

right external jugular vein

right internal carotid artery

right external carotid artery

right internal jugular vein

ductus caroticus

right systemic artery

right systemic arch

right pulmonary artery

pharynx

right superior vena cava

subclavian vein

right azygos vein

subclavian artery

brachial nerve

dorsal aorta

right pulmonary artery

right lung

oesophagus

artery to body wall

oesophagus

left carotid arch

left systemic arch

left internal jugular vein

pulmonary arch

left auricle

subclavian vein

ventricle

inferior vena cava

left lung

liver

Fig. 23. Lizard dissected to show the heart and arterial system. The pectoral girdle has been cut and laid back (cut surface cross-hatched) and the pericardium and right auricle removed.

into the left auricle. Immediately dorsal to the pulmonary vein lies the pulmonary artery.

The heart and arterial arches (Fig. 23)

Remove both auricles and observe the three arteries twisted together to form what looks like a truncus arteriosus. There is in reality no truncus arteriosus, the division of the arteries being carried right back to the ventricle itself.

The pulmonary artery is single and arises on the right-hand side of the ventricle. It passes over obliquely to the left side and then curls dorsally to reach the anterior end of the pericardium. Here it divides into two pulmonary arteries which run backwards alongside the trachea on either side of the lungs. At first the pulmonary artery runs close to the pulmonary vein, but on reaching the lung the artery passes along its outer border whereas the vein runs along the inner border of the lung.

The left systemic arch arises from the ventricle slightly to the left of the pulmonary artery. The left systemic arch in the dissection can be seen lying along the right side of the pulmonary artery. It twists over the pulmonary artery towards the left side and runs outwards and forwards towards the angle of the jaw. The remaining part of the left systemic arch can best be dissected out by starting in the body cavity. Find the dorsal aorta and trace it forwards until it divides into the two systemic arches. Follow the left systemic arch forwards to the point where it has been left.

The right systemic arch arises from the ventricle slightly to the left of the origin of the left systemic. The right systemic curls through the pericardial wall outwards towards the angle of the jaw. Find the posterior part of this systemic arch in the same way as was done on the left side.

Just as the right systemic arch passes away from the left, two carotid arteries arise from it close together. Follow the carotid arteries outwards. They run parallel with the systemic arches towards the angle of the jaw. About $\frac{1}{4}$ inch away from the origin of the carotids the small paired external carotid artery is given off on either side. The external carotid artery supplies the floor of the mouth. Follow the carotid artery forwards beyond the origin of the external carotid towards the base of the skull where it forms the internal carotid artery.

About $\frac{1}{8}$ inch beyond the origin of the external carotid artery the systemic and carotid arches are united by an open vessel, the ductus caroticus (*ductus Botalli*). The ductus caroticus is about the same size as the systemic arch. It represents a portion of the paired dorsal aortae.

Both subclavian arteries arise from the right systemic arch. Find the right subclavian lying in the middle of the brachial plexus a short distance behind the subclavian vein. Follow it inwards towards the middle line. In order to see the junction of the subclavian and systemic arches the strip of muscle lying at the side of the systemic arch must be removed. The subclavian artery unites with the systemic arch a short distance in front of the point of union of the two systemics. Find the left subclavian artery arising from the right systemic, opposite the origin of the right subclavian.

The pharynx and lungs (Fig. 23)

Cut open one of the lungs along its entire length. Note that the lung is vascular and alveolated throughout. The wall in front is thick and spongy while posteriorly it is thinner.

Each lung communicates with the very short bronchus. The opening of the bronchus into the lung is a short distance behind the anterior end of the lung. The two bronchi unite to form the long trachea. Both the trachea and the bronchi are stiffened by complete cartilaginous rings. Follow the trachea forwards and find the opening into the mouth, the glottis. The walls of the anterior end of the trachea just before it opens into the mouth form a round prominence, the larynx. The larynx is stiffened by cartilage derived from the remains of the gill arches.

The hyoid covers the ventral surface of the anterior portion of the trachea. It is a median plate from which extend backwards two long horns.

The thymus gland will be seen projecting forwards between the origin of the two carotids. The major portion of it lies dorsal to and is obscured by the origin of the carotids. The thyroid gland lies further forwards across the trachea about $\frac{1}{3}$ inch in front of the thymus. It is crescentic in shape.

Remove the lower jaw by cutting through the angles. Note the internal nares at the anterior end of the oral cavity.

The teeth are very small and are borne on the premaxilla, maxilla, and the palatine bones.

The Eustachian tube is a very large recess lined by pigmented membrane. Hold the animal up to the light, and note the transparent tympanic membrane closing the Eustachian tube to the exterior.

The eye

Remove the eye from the orbit. The relationships of the different layers will be found to be essentially the same as in the skate.

8 *The grass snake*

Tropidonotus natrix

External features OBSERVE the elongated shape of the body and note the entire absence of limbs. This elongation of the body is produced by drawing out the trunk region, the tail region behind the cloaca is no more developed than it is in Lacertilia.

Find the cloaca a short distance from the posterior end of the body. It is a transverse slit, bounded in front by two scales which fold over it and make the opening rather inconspicuous.

The general surface of the body is covered by epidermal scales arranged in a regular pattern. The scales on the dorsal surface are small except in the head region where they are large and fit closely to the skull. As in the lizard, these large scales on the head do not correspond in arrangement with the underlying bones of the skull.

The scales on the ventral surface from the throat to the cloaca are large, overlapping posteriorly and extending completely across the ventral surface. The ribs are connected to these scales and some snakes move by raising these scales and depressing them one after another in succession from the front backwards. This enables the snake to glide over the ground without executing serpentine movements. The scales behind the cloaca and the one immediately in front of it on the ventral surface are paired and do not stretch completely across the ventral surface. The mouth is terminal. Open the mouth and feel the teeth which are on the premaxilla, maxilla, pterygoid, and palatine bones in the upper jaw and on the dentary in the lower jaw. The internal nares can be seen in the roof of the mouth. In the floor of the mouth note the median prominence which is the larynx. Find the glottis in the middle of this prominence and insert down it a probe. Feel the cartilages supporting the larynx on either side. Just anterior to the larynx will usually be seen the forked end of the tongue protruding from its pouch. In some specimens it will be possible to catch hold of the tongue and pull it forwards to its extreme extent which is about $1\frac{1}{2}$ or 2 inches.

The external nares are situated on either side of the snout. The eyes are on either side of the head. The eyelids are permanently fused over the eyes but form transparent 'spectacles'.

General directions for dissection

Cut the animal open by a median ventral incision extending from the lower jaw to the cloaca. Pin the sides of the body down so as to expose the viscera.

It will be best first to obtain a general idea of the arrangement of the contents of the body cavity by identifying certain of the more conspicuous structures, and then to make a complete investigation by dissecting the animal in three portions.

At the anterior end a few inches behind the head note the heart. Immediately behind the heart is a conspicuous vessel, the inferior vena cava. Alongside the vena cava will be seen, on the left-hand side, the oesophagus, and on the right the lung. The lung extends a long way backwards. Only the anterior portion is respiratory and has thick walls, while the posterior portion has very thin transparent walls. About 2 inches behind the heart is the liver, an elongated brown body, not divided into lobes. It extends from one-third to half way down the body. The inferior vena cava runs along its ventral surface and continues backwards at the posterior end. Behind the liver the rest of the viscera are covered by the fat body; lying in the middle of the fat body is the anterior abdominal vein. About 2 inches behind the posterior end of the liver will be seen the large gall-bladder, greenish in colour. Immediately behind the gall-bladder lies the pancreas. From this point backwards the intestine is much coiled. The kidneys are elongated lobular structures lying at the sides of the alimentary canal, the right kidney being more anterior in position than the left one. The posterior portion of the body cavity contents should be handled as little as possible before dissection of the urinogenital system is undertaken.

The anterior third of the body (Fig. 24)

At the anterior end of the body observe the very long muscular pouch in which lies the tongue. Raise this pouch and pin it aside, thus exposing the trachea. A certain amount of fat may be found on either side of the trachea. This should be carefully picked off. Just in front of the heart lying on the ventral side of the trachea is the small round thyroid gland.

Heart and veins

Carefully remove the pericardium from the heart. Note the general similarity of the heart to that of the lizard. The right auricle is larger than the left which lies dorsal to the arteries originating from the ventricle. Turn the heart to the left side and observe the sinus venosus entering the right auricle. At the posterior end of the sinus the large inferior vena cava enters. Anteriorly the sinus venosus receives two superior venae cavae.

Follow the right vena cava forwards. Almost immediately after leaving the sinus vensosus it receives the azygos vein. The azygos drains the intercostal muscles on both sides, as in the lizard. In front of the auricle the right vena cava divides into the jugular and anterior azygos veins. The jugular vein is a large vessel running ventrally beside the trachea and draining both sides of the body in this region. The anterior azygos runs forwards dorsal to the jugular and drains the body wall in the neck region and the contents of the cranium.

The left superior vena cava is formed by the jugular only. It runs forwards to the head and drains blood from the same region as the right jugular on the other side.

Turn the heart over to the left side and carefully remove the dorsal wall of the pericardium. Find the pulmonary vein lying close alongside the dorsal wall of the inferior vena cava. Trace it backwards to the lung and forwards where it can be seen to run into the left auricle.

Arteries in the region of the heart

The arterial arches arise directly from the ventricle as in the lizard and twist round each other in a similar manner.

The pulmonary arch arises on the right-hand side of the ventral surface of the ventricle. It curls across to the left and soon disappears from view. Turn the heart over to the left and trace the course of the pulmonary arch, which is single and runs in the cleft between the two auricles. Follow the pulmonary artery along the outer border of the lung. Farther back along the lung it lies close to the inferior vena cava.

The next arch is the left systemic. This runs forwards as far as the anterior border of the left auricle, where it turns dorsally and then curls back along the dorsal side of the body. After running along the dorsal side for about an inch or so it unites with the systemic arch of the other side to form the dorsal aorta. This union will be found a little way behind the level of the posterior border of the heart.

The right systemic arch is obscured at its origin by the pulmonary and left systemic arches. Carefully dissect the left systemic arch away from the right so as to separate the two. The right systemic arch curls round and runs backwards along the dorsal side of the body in the same manner as the left.

Just before the right systemic arch curls dorsally it gives off the common carotid artery which runs forwards alongside the

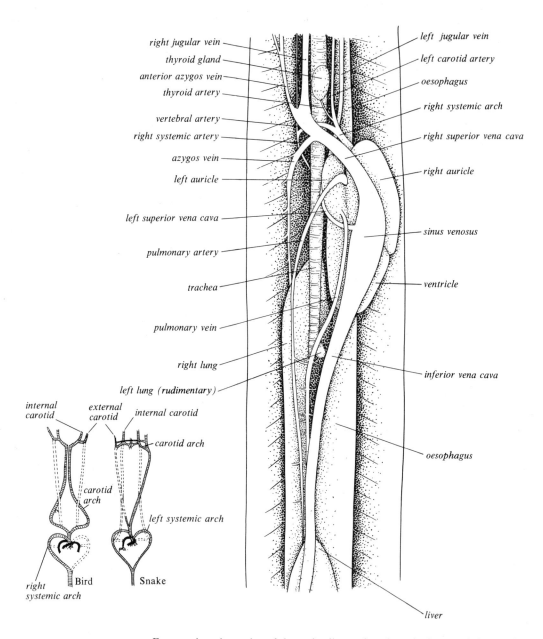

right jugular vein

thyroid gland

anterior azygos vein

thyroid artery

vertebral artery

right systemic artery

azygos vein

left auricle

left superior vena cava

pulmonary artery

trachea

pulmonary vein

right lung

left lung (rudimentary)

left jugular vein

left carotid artery

oesophagus

right systemic arch

right superior vena cava

right auricle

sinus venosus

ventricle

inferior vena cava

oesophagus

internal carotid

external carotid

internal carotid

carotid arch

carotid arch

left systemic arch

right systemic arch

Bird

Snake

liver

FIG. 24. Anterior region of the snake dissected to show the heart and the arterial and venous system round it. The pericardium has been removed and the heart twisted over to the left so as to expose its dorsal side. The diagrams show the manner of formation of the carotid arteries in the long neck of the snake and the bird.

trachea close to the left internal jugular vein as the left carotid artery. On reaching the angle of the jaw on the left-hand side it branches into two, the internal carotid which runs dorsally and enters the skull supplying the brain, and the external carotid which runs along the floor of the mouth. The right-hand side of the head and skull is supplied by a vessel that runs from the left external carotid across to the right side passing near the symphysis of the jaw. The internal carotid similarly sends a branch across to the other side close to its origin.

The right carotid artery is minute and supplies the thyroid gland only, arising from the common carotid near the systemic arch.[1]

About ¼ inch from the origin of the carotid artery from the right systemic arch and almost exactly at the point where this arch reaches the dorsal wall there arises the vertebral artery. It runs forwards supplying the musculature in the mid-dorsal line. It also gives small branches to the oesophagus. The right systemic arch beyond this point is much smaller than the left.

Trace the dorsal aorta backwards from the point where it is formed by the union of the two systemic arches. In doing this notice the oesophagus, which merges gradually into the stomach and lies as a rule on the left side of the animal, while the lung lies on the right with the liver in between. On its course backwards the dorsal aorta gives off many small branches to the oesophagus and the musculature of the body wall.

The eye

Remove the eye from the orbit. The relationships of the different layers will be found to be essentially the same as in the skate.

Middle third of the body
(Fig. 25)

Raise the liver and observe on its dorsal side the hepatic portal vein which runs in a similar position to the inferior vena cava on its ventral surface. At the anterior end of the liver note that the hepatic portal vein does not proceed any farther forwards. It breaks up into capillaries in the liver. At the posterior end of the liver note the two large entering vessels. The ventral of these is the inferior vena cava and the lower (dorsal) one is the hepatic portal vein.

About an inch or so behind the posterior border of the liver will be seen, after the removal of the fat, the gall-bladder filled

---

[1] In the snake the neck elongates and draws the systemic and carotid arches apart, the carotid arch remaining close to the head. The paired dorsal and ventral aortae between these arches thus become elongated. The external and internal carotids then unite across the middle line. The elongated paired ductus caroticus (paired dorsal aorta) then disappears on either side and the single carotid artery along the neck thus represents an elongated part of the right ventral aorta. (See Fig. 24.)

with green bile. In removing the fat be careful not to damage the anterior abdominal vein. Near the posterior end of the liver the very long oesophagus merges into the short stomach which extends back to the level of the gall-bladder. Just below this level will be seen a marked constriction of the alimentary canal forming the pylorus. Behind the pylorus the intestine is coiled.

Immediately behind the gall-bladder lies the pancreas. It is a compact body of about the same size as the gall-bladder. The spleen is very small and is closely applied to the anterior border of the pancreas. About half an inch from the posterior border of the pancreas the anterior abdominal vein enters the hepatic portal vein.

Carefully remove the connective tissue round the gall-bladder so as to free it and expose its ducts. The cystic duct arises at the anterior end of the gall-bladder. It runs along its border to the pancreas. Just before entering the pancreas, the cystic duct will be seen to receive a hepatic duct from the liver. This duct follows the course of the inferior vena cava and the hepatic portal vein, lying between them and entering the liver at its extreme posterior border. The common bile-duct passes through the middle of the pancreas receiving the pancreatic ducts. The duct leaving the pancreas is extremely short and passes from the dorsal surface of the pancreas into the ventral side of the intestine.

Separate the inferior vena cava from the hepatic portal vein and follow them both backwards. The hepatic portal vein runs along the dorsal surface of the stomach and intestine receiving branches from both and from the body wall. Just behind the pancreas, as has already been described, it receives the anterior abdominal vein. The inferior vena cava runs straight backwards without receiving any branches until it reaches the kidneys.

The dorsal aorta gives off numerous branches on its course backwards. Some of these are very small. They supply the muscles of the body wall and viscera. It will only be possible to enumerate the larger of these vessels. On passing the stomach the dorsal aorta gives off several small gastric branches. Opposite the gall-bladder the large lienogastric artery is given off, supplying the stomach and spleen. Behind the pancreas arises the superior mesenteric which supplies the pancreas and a considerable length of the anterior portion of the intestine.

The single lung extends backwards well behind the level of the pancreas. Follow out the entire length of the lung and cut it open so as to expose its inner walls. Anteriorly the walls are thick and spongy, containing numerous small alveoli, while

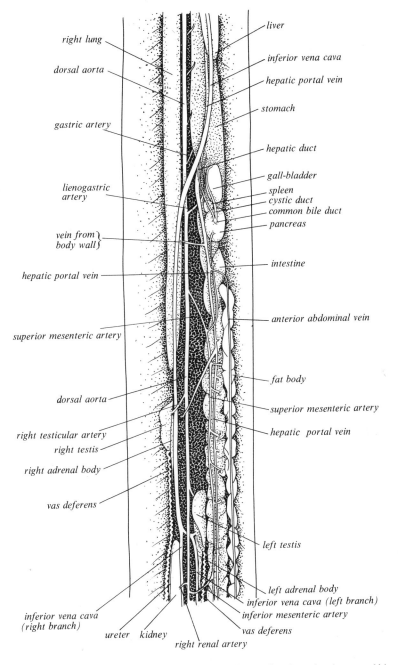

right lung

dorsal aorta

gastric artery

lienogastric artery

vein from body wall

hepatic portal vein

superior mesenteric artery

dorsal aorta

right testicular artery

right testis

right adrenal body

vas deferens

inferior vena cava (right branch)

ureter   kidney

right renal artery

liver

inferior vena cava

hepatic portal vein

stomach

hepatic duct

gall-bladder

spleen

cystic duct

common bile duct

pancreas

intestine

anterior abdominal vein

fat body

superior mesenteric artery

hepatic portal vein

left testis

left adrenal body

inferior vena cava (left branch)

inferior mesenteric artery

vas deferens

FIG. 25. Middle region of the snake dissected to show the viscera and blood-vessels. The connective tissue only has been removed.

posteriorly the walls are thin and much smoother internally. Examine the junction of the trachea with the lung. Note that the alveolar tissue of the lung extends forward beyond the level of the opening of trachea into the lung. Occasionally a rudimentary left lung may be seen as a slight bulge projecting from this region on the left side.

Posterior third of the body

Remove the fat body from the remaining portion of the body cavity. Follow the coiled gut backwards. Do not stretch the gut far from the dorsal body wall or the arteries leading to it from the dorsal aorta will be broken. About 2 or 3 inches in front of the cloaca the intestine dilates into a thin-walled rectum. The rectum is a straight tube with thinner walls near the opening into the cloaca. This portion serves to contain the faeces and the urinary excreta which are ejected together in a semi-fluid condition. There is no separate urinary bladder.

Slit open the cloaca and note the transverse septum in the same position as in the lizard.

Urinogenital system
(Fig. 25)

The paired testes or ovaries lie suspended from the dorsal body wall close to the aorta. The right gonad may lie a considerable distance in front of the left. The position of the paired kidneys has already been noticed.

Male. The testes are small compact bodies about $\frac{3}{4}$ inch in length. They are attached to the dorsal body wall by a mesentery (the mesorchium). Outside this mesentery lies the epididymis (reduced mesonephros). The Wolffian duct forms the coiled vas deferens. On the right-hand side this runs backwards alongside the dorsal aorta and then follows the ventral border of the right kidney. On the left-hand side, the vas deferens runs backwards alongside the left branch of the vena cava and then, as on the right side, along the ventral border of the left kidney. Both vasa deferentia then continue backwards close to the rectum and open into the dorsal wall of the posterior chamber of the cloaca by a median pore.

Attached on either side of the cloaca and projecting forwards into the body cavity are the paired corpora cavernosa that constitute the penes. These, as in the lizard, are vascular structures which are dilated by the pressure of the blood-vessels in them and extruded through the cloacal aperture.

The metanephric kidneys are brown lobulated structures about $2\frac{1}{2}$ inches in length. The right-hand kidney is situated some distance in front of the left. The ureters lie along the ventral borders of the kidneys close to the vasa deferentia. The ureters pass backwards alongside the vasa deferentia and open with them into the posterior chamber of the cloaca. The adrenal bodies are slender elongated masses of yellowish tissue lying on the inner side of the gonads, between these organs and the branches of the inferior vena cava. In the male they extend backwards beyond the posterior border of the testis. *Female.* The ovaries vary enormously in size according to the condition of the animal. Large ova may, as a rule, be seen projecting from the walls.

The oviduct is pleated and lies along the outer side of the ovary. The internal opening of the oviduct lies a little way in front of the ovary. Both oviducts pass backwards and open into the posterior chamber of the cloaca by a median dorsal opening which is raised on a papilla.

The kidneys and the ureters are similar to those of the male. The ureters open together with the oviducts into the cloaca.

Vascular system of the posterior region (Fig. 25)

Follow the posterior vena cava backwards. Just in front of the right kidney it divides into two branches, which arise along the dorsal sides of the kidneys receiving blood from them.

Follow the hepatic portal vein backwards. It lies along the dorsal side of the intestine and the rectum receiving branches from the alimentary canal and the body wall. The origin of this vessel is difficult to see. It is formed by the union of two vessels arising from the renal portal veins in the region of the cloaca.

Follow the dorsal aorta backwards. The order in which the arteries arise from it to supply the various organs in the body cavity is subject to variation. The superior mesenteric artery has already been identified. The order in which the arteries posterior to this usually arise is as follows: the right gonadial artery, the first inferior mesenteric, the anterior right renal, the left gonadial, the anterior left renal, the second inferior mesenteric, the median right renal, the posterior right renal, the third inferior mesenteric, the median left renal, the fourth inferior mesenteric, and the posterior left renal. In the male the first inferior mesenteric artery arises posteriorly to the anterior left renal artery, whereas in the female it is anterior to the anterior right renal.

The anterior abdominal vein arises posteriorly in the fat body. There are of course no pelvic veins, so the anterior abdominal vein is without its usual source of supply.

The renal portal veins arise by a bifurcation of the caudal vein in the cloacal region. They pass forwards close to the ureters and lie along the ventral borders of the kidneys.

Cut vertically downwards through the muscles behind the cloaca and find the caudal vein and the caudal artery lying side by side just ventral to the vertebral column.

9 *The pigeon*

Columba livia

External features

General

THE general shape of the body of a bird is entirely different from any vertebrate hitherto studied. The modifications are due to the adoption of bipedal gait and the habit of flight. The body is carried well off the ground as the animal stands on its two hind limbs. When the bird is standing upright the centre of gravity tends to be well in front of the two hip joints but over the toes.

The head is separated from the trunk by a well marked neck. The trunk is deep dorsoventrally. The tail region is very short.

The mouth is situated at the anterior end of the body. The horny covering of the jaws forms the beak.

The external nares lie at the posterior end of the upper beak. Just above the external nares is a swollen patch of skin, usually white in colour, called the cere.

Immediately behind the mouth is the eye. This is provided with two eyelids; there is also a flap of transparent skin which can be drawn over the eye from the upper anterior border, the nictitating membrane. The nictitating membrane can easily be pulled down by a pair of forceps.

The external opening of the ear lies about half an inch below the posterior border of the eye. The opening is easily seen by raising the feathers in this region.

Stretch out one of the wings. The bones of the wings can be identified by feeling for them through the skin. A comparison should be made with the skeleton of the wing prepared with the feathers attached.

The surface of the wing is greatly increased by the feathers which are attached to the hand and forearm. These feathers are composed of large quills, the remiges, divisible into the primaries and secondaries. The primaries are the longer feathers attached to the fused second and third digits. The shorter secondaries are attached to the ulnar bone of the fore-arm. The primaries and the secondaries merge into each other so gradually in the region of the wrist that they cannot be distinguished by their size alone. It will probably be necessary to look at the prepared skeleton in order to be certain of the point of division. At their bases the primaries and secondaries are overlain by the wing coverts.

The bastard wing is a small group of feathers attached to the pollex.

The long tail feathers, or rectrices, are of the same nature as the primary and secondary feathers of the wing. They are also overlain at their bases by tail coverts.

Raise the tail coverts and immediately behind the insertion of the rectrices will be seen a prominent papilla at the tip of which is the opening of the oil gland. It secretes a substance which is spread over the feathers when the animal preens itself.

The cloaca lies at the base of the tail on the ventral surface. As in Reptiles it opens by a transverse slit.

The leg is bare of feathers below the tarsal joint and the skin is here covered with horny epidermal scales. There are four digits (fifth missing). The hallux is directed backwards and is of assistance in grasping boughs when the animal perches.

Feathers

Except for the feet the surface of the body is everywhere covered with feathers. These are specially modified epidermal structures and they are one of the most characteristic features of the bird. The feathers are arranged in regular tracts, called pterylae. These pterylae are not so distinct in the pigeon as in some other birds and can only be distinguished with difficulty.

STRUCTURE OF A FEATHER. Pull out one of the large primaries from the wing. The feather consists of a central stem (a quill or rachis) bearing on either side the vanes. The rachis consists of a stiff solid portion between the vanes and a hollow portion below them. At the base of the rachis is a small hole, the inferior umbilicus, leading into the hollow portion of the rachis, and at the junction of the solid rachis and the hollow base there is another small hole, the superior umbilicus.

The vane is made up of a large number of barbs on either side of the rachis. These barbs in turn have a number of barbules on either side. The barbules are provided with hooks interlocking with the barbules of the barbs on either side. In order to see the barbs and barbules it will be necessary to cut off a piece of the feather, mount it in alcohol, and examine it under the microscope.

VARIETIES OF FEATHERS. The contour feathers cover the

general surface of the body. They are smaller than the wing or tail feathers, and the hooks on the barbules are not so well developed, especially at the basal portion which has a downy appearance.

The down feathers are found chiefly on the breast. In these the barbules and hooks are hardly developed at all, so that the feather has no stiff vane.

The filoplumes are minute feathers resembling hairs with a tuft of barbs at the tip. They are easily seen remaining in the skin after the bird is plucked.

General directions for dissection

Skinning, and inflation of the air sacs

First pluck the feathers from the ventral surface of the body and neck. In plucking the feathers from the neck region be careful not to tear the skin.

Remove the skin from the ventral surface of the thorax and abdomen. Commence by making an incision in the mid-ventral line and pulling the skin away on either side, thus exposing the large pectoral muscles. Cut through the skin in the mid-ventral line along the neck and lay it back on either side. When doing this it will be necessary also to separate the walls of the crop from the skin to which it adheres very closely. Avoid, if possible, breaking the crop. Expose the trachea in the middle of the neck, and separate a small portion of it from the surrounding tissue afterwards placing a ligature round it ready for tying. Just anterior to this ligature make a small slit in the trachea and insert a blowpipe. Blow down the pipe and observe that the abdomen is distended and the sternum raised. Draw the ligature tight—not too tight, or the trachea will be severed—with the abdomen fully distended and withdraw the blowpipe. The effect of this is to distend the air sacs with air and to make them more easily seen in the subsequent dissection.

Separate the crop from the pectoral muscles by pulling it away with the fingers. In the space between the furcula (fused clavicles) observe the front portion of the median interclavicular air sac. The crop must be laid back so as to expose the dorsal end of each clavicle.

Muscles of the pectoral girdle and wing
(Figs. 26(a), (b), 27

Before beginning to dissect the muscles, examine either a mounted skeleton of the pigeon or separated parts of the skeleton. Read pp. 249–254 and observe the shapes of the bones that comprise the girdles and limbs. Notice the lightness of the skeleton; this is the result of the presence of air spaces within many of the bones.

NOTE. The description given here of the muscles refers only to the pigeon. There is a basic similarity between the muscles of different species of birds, but there are many differences in detail. If a bird other than a pigeon is dissected, this description should be generally applicable, but certain muscles may be completely absent, others may be of quite different form, and muscles may be found that are not present in the pigeon.

Remove the skin over the side of the sternum and back and notice several slips of *cutaneus* muscle inserting in the skin and arising from fascia over other muscles. Carefully remove the skin from the dorsal and ventral surfaces of the wing, taking care not to injure the *patagial tendon* that lies along its anterior border. Clear away the superficial fat and connective tissue over the muscles.

Place the bird on its back and with a sharp scalpel cut along the edge of the keel of the sternum which can be seen in the mid-ventral line as a narrow bone. Separate the muscle from the keel, draw it outwards, and find the fascia separating the superficial *pectoralis* from the deep *supracoracoideus*. Detach the *pectoralis* from the keel and then cut it from its origin on the edge of the clavicle and the membrane between the clavicle and coracoid, starting in the mid-line and working laterally and exposing the axillary air sac. Then separate the pectoralis from the surface of the supracoracoideus and cut the former from its origin on the outer edge of the sternum, beginning at the posterior end (Fig. 27, p. 105). This will expose an air sac between the two muscles with the pectoral arteries and veins visible through it. Ligature these blood-vessels, which supply the two muscles, and then cut through them distal to the heart. Detach the pectoralis from the rest of its origin and observe its insertion on the humeral crest; pull on this muscle, observe that it lowers the wing. Then reflect the muscle forwards.

Detach the *supracoracoideus* from the surface and keel of the sternum and from the membrane between the coracoid and clavicle beginning with its caudal end. Observe that the muscle converges on its dorsal surface to a tendon that disappears into the foramen triosseum at the junction of the coracoid, clavicle, and scapula. Support the bird in an upright position in one hand and pull on the supracoracoideus and note that the wing is raised. Pull alternately on the pectoralis and supracoracoideus, lowering and raising the wing alternately. Cut away the pectoralis from its insertion on the humeral crest (deltoid ridge).

Place the bird on its ventral surface and examine the muscles of the back and shoulder. Find the *latissimus dorsi*, a thin slip of muscle arising from vertebral spines and inserted

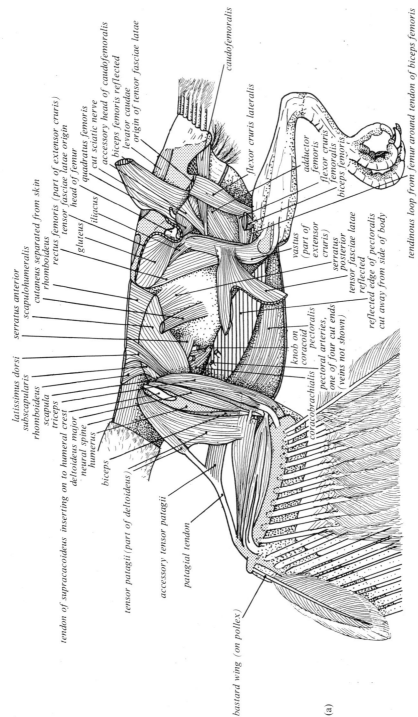

latissimus dorsi
subscapularis
rhomboideus
scapula
triceps
neural spine
deltoideus major
humerus
biceps

tendon of supracoracoideus inserting on to humeral crest

serratus anterior
scapulohumeralis
cutaneus separated from skin
rhomboideus
rectus femoris (part of extensor cruris)
tensor fasciae latae origin
head of femur
gluteus
quadratus femoris
iliacus
accessory head of caudofemoralis
biceps femoris reflected
levator caudae
cut sciatic nerve
origin of tensor fasciae latae

caudofemoralis

flexor cruris lateralis

adductor
femoris
flexor cruris
femoralis
biceps femoris

vastus
(part of
extensor
cruris)
serratus
posterior
tensor fasciae latae
reflected
reflected edge of pectoralis
cut away from side of body

knob on
coracoid
pectoralis
pectoral arteries,
one of four cut ends
(veins not shown)
coracobrachialis

tendinous loop from femur around tendon of biceps femoris

tensor patagii (part of deltoideus)

accessory tensor patagii

patagial tendon

bastard wing (on pollex)

(a)

Fig. 26(a). Pigeon. Lateral view showing the superficial muscles and the clenched position of the toes consequent on the flexure of the ankle. The skin, fat and connective tissue have been removed, tendons are white, cut surfaces cross-hatched, and bone exposed or nearly so is shown by mechanical tint. The cutaneus is cut away from the skin, tensor fasciae latae and biceps femoris are reflected, and the lateral edge of pectoralis is reflected from the side of the body exposing the cut pectoral arteries and coracobrachialis.

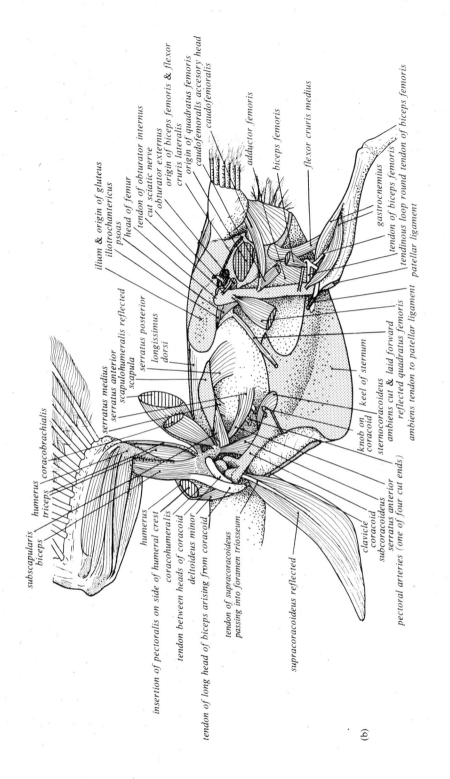

Fig. 26(b). Pigeon. Lateral view showing the deeper muscles and using the same conventions as in the preceding figure. The following have been removed: pectoralis, cutaneus, rhomboideus, latissimus dorsi, tensor patagii, deltoideus major, tensor fasciae latae, biceps femoris, gluteus, rectus femoris, vastus, flexor cruris lateralis and femoris. The following have been reflected: supracoracoideus, coracobrachialis, scapulohumeralis, ambiens, quadratus femoris, accessory head of caudofemoralis, and gastrocnemius (to show insertion of biceps femoris).

(b)

subscapularis
biceps
humerus
triceps coracobrachialis

humerus
coracohumeralis
tendon between heads of coracoid
deltoideus minor

insertion of pectoralis on side of humeral crest

tendon of long head of biceps arising from coracoid

tendon of supracoracoideus
passing into foramen triosseum

pectoral arteries (one of four cut ends)

supracoracoideus reflected

clavicle
coracoid
subcoracoideus
serratus anterior

serratus medius
serratus anterior
scapulohumeralis reflected
scapula

serratus posterior
longissimus
dorsi

ilium & origin of gluteus
iliotrochantericus
psoas
head of femur
tendon of obturator internus
cut sciatic nerve
obturator externus
origin of biceps femoris & flexor
cruris lateralis
origin of quadratus femoris
caudofemoralis accesory head
caudofemoralis

adductor femoris

biceps femoris

flexor cruris medius

gastrocnemius
tendon of biceps femoris
tendinous loop round tendon of biceps femoris
patellar ligament

knob on
coracoid
keel of sternum
sternocoracoideus
ambiens cut & laid forward
reflected quadratus femoris
ambiens tendon to patellar ligament

on to the shaft of the humerus. Pull on this muscle and observe that the humerus is raised and retracted; cut it away. Find the *rhomboideus* muscles forming a thin sheet arising from the vertebral spines and inserted on the dorsal edge of the scapula, which they raise. Remove them and raise the scapula in order to see below it, inserted on its inner surface, three groups of *serratus* muscles arising from the ribs. These should be investigated later.

Note the large *scapulohumeralis* arising from the upper surface and lower edge of the posterior part of the scapula and inserted on the shaft of the humerus, which it retracts. Cut this muscle away. Now investigate the *deltoideus*, which can be divided into the following parts. (1) Tensor patagii arises from the coracoscapular junction and is inserted on to the patagial tendon and also on to fascia over the radio-ulna. Pull on it and observe that the humerus is protracted and the elbow joint is flexed. Extend the elbow joint and observe that the wrist simultaneously extends through the action of the patagial tendon. Note the accessory tensor patagii arising from the surface of the biceps muscle and also inserted on the patagial tendon. Remove both these muscles and the tendon. (2) Deltoideus major arises from the anterior end of the scapula and inserts on the shaft of the humerus which it protracts and raises. (3) Deltoideus minor is a short muscle below tensor patagii and inserted on the humeral crest (Fig. 27); it protracts the humerus. Cut away deltoideus major and minor and find the scapular head of the *triceps* immediately posterior to deltoideus major and inserted on the ulna. This extends the elbow joint. Remove it. There is now exposed the tendon of the *supracoracoideus* which emerges from the foramen triosseum and is inserted on the head of the humerus. Pull on the muscular part of the supracoracoideus (which is on the ventral side) to show that this tendon is connected to it and then cut away the muscle.

Turn the bird over on to its back and find the *coracobrachialis* arising mainly from the lateral surface of the coracoid and inserted on the posterior surface of the humerus which it retracts. Cut this muscle away. Examine the articulation between the coracoid and the sternum and find the small *sternocoracoideus*, which connects the two bones. Note the *subcoracoideus* arising from the dorsal surface of the coracoid and the anterior end of the clavicle and inserted on the head of the humerus together with the *subscapularis* which arises from the anterior end of the scapula. These two muscles can retract the humerus and the former also rotates it. The subscapularis is penetrated by the tendon of the anterior *serratus* muscle (Fig. 27). Remove the subcoracoideus and

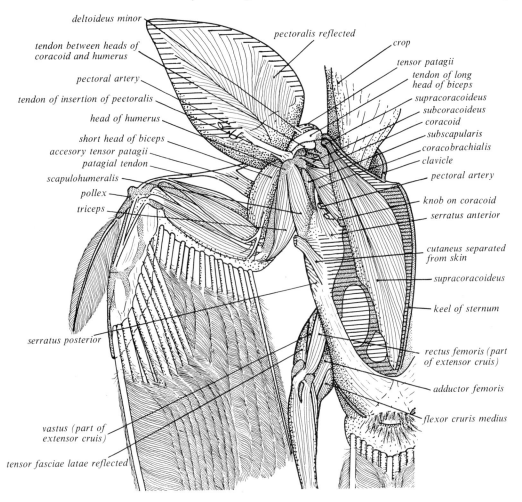

deltoideus minor

tendon between heads of coracoid and humerus

pectoral artery

tendon of insertion of pectoralis

head of humerus

short head of biceps

accesory tensor patagii

patagial tendon

scapulohumeralis

pollex

triceps

serratus posterior

vastus (part of extensor cruis)

tensor fasciae latae reflected

pectoralis reflected

crop

tensor patagii

tendon of long head of biceps

supracoracoideus

subcoracoideus

coracoid

subscapularis

coracobrachialis

clavicle

pectoral artery

knob on coracoid

serratus anterior

cutaneus separated from skin

supracoracoideus

keel of sternum

rectus femoris (part of extensor cruis)

adductor femoris

flexor cruris medius

FIG. 27. Pigeon, ventral view using the same conventions as in Fig. 26(a). The cutaneus and pectoralis have been reflected.

subscapularis and examine the three serratus muscles in more detail.

Find the tendon of the long head of the *biceps*, the flexor of the elbow joint. This muscle arises from the anterior end of the coracoid. Cut the muscle away and note the short head of the biceps arising from the head of the humerus. The two heads join and inert on the radio-ulna. Remove the biceps. There is now exposed the *coracohumeralis*, a short protractor of the humerus, arising from the anterior border of the glenoid and inserted on the base of the humeral crest (Fig. 26(b)).

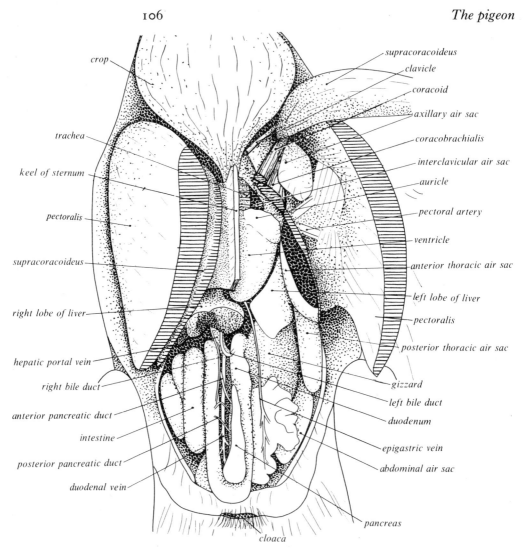

FIG. 28. Pigeon. On the right side the muscles have been cut free from the keel of the sternum but not yet turned back, and on the left side the pectoralis and supracoracoideus muscles have been laid back and the sternum removed. The air sacs are shown on the left side, and part of the wall of the anterior thoracic air sac has been cut in opening the body. The great omentum has been removed from the right side of the abdomen, exposing the intestine. The right lobe of the liver is turned up exposing the bile ducts, and the duodenal loop has been pulled out.

Remove this muscle and then examine the head of the humerus and investigate the ligaments which control its movements in the glenoid cavity.

The scapular head of the *triceps* has already been cut away. Find the two short heads of the muscle, arising on the posterior surface of the humerus and cut them away. On either surface of the radio-ulna, observe extensors, flexors, and rotators of the wrist and short flexors and rotators of the elbow joint. Identify the functions of these muscles by clearing each muscle of connective tissue, pulling on it and observing the result. Cut each muscle away when its actions have been investigated and thus expose the deeper muscles. Find the muscles that operate the bastard wing (first digit) and examine the muscles and tendons distal to the wrist joint. Dissect away connective tissue to expose the base of the primary and secondary feathers and observe how they are attached to the wing bones and the ligaments that control their movements. Flex the elbow joint and note that the wrist is flexed simultaneously by the action of these ligaments, which arise on the humerus and end on the phalanges.

Contents of the body cavity
(Figs. 28, 29)

The sternum can now be removed by cutting through its posterior attachment to the body wall. Laterally the cut must be made through the attachment of the ribs to the sternum and through the ventral part of the coracoids, passing ventrally to the pectoral arteries and veins. Great care must be taken in cutting away the sternum in order to avoid damaging the liver and heart, lying immediately below the sternum, and the pectoral arteries and veins already mentioned. Anteriorly the sternum can easily be separated from the fused ends of the clavicles. The sternum can now be lifted off. Cut through the ventral body wall posteriorly in the middle line as far as the cloaca.

Preliminary inspection

At the anterior end of the body cavity lies the heart enclosed in a thin pericardial membrane. At the anterior end of the heart lie the two auricles. Immediately in front of the auricles lies the interclavicular air sac. Through the transparent walls of this air sac may be seen the posterior end of the trachea. Immediately behind the heart lies the liver, divisible into two main lobes, the right one being the larger of the two. Behind the liver the viscera are covered by a flap of connective tissue, the great omentum, which is usually heavily loaded with fat.

Air sacs

The interclavicular air sac has already been seen. Gently push the viscera away from the body wall on one side and observe the air sacs. The most anterior of these lying ventral to the lung is the anterior thoracic air sac. Part of its ventral wall will have been removed with the sternum and the pink lung may be seen through its upper dorsal walls. Behind and partly dorsal to the anterior thoracic air sac lies the posterior thoracic air sac. The walls of this sac are usually left intact. Behind the posterior air sac lies the abdominal air sac in the hinder part of the body cavity. These three air sacs are paired and communicate separately with certain of the bronchial tubes of the lungs. Some of these air sacs are prolonged into cavities in the bones.

Alimentary canal and glands

Separate the two lobes of the liver. The gizzard will be seen immediately behind the smaller left-hand lobe. Running from the great omentum forwards past the posterior border of the gizzard along its right-hand side will be seen the epigastric vein. This passes forwards between the two lobes of the liver and opens into the left hepatic vein.

Carefully separate the great omentum from the underlying structures, turning it on to the right-hand side, so exposing the viscera. In the middle line will be seen the pancreas lying between the two limbs of the duodenum. Pull out the duodenum and stretch it backwards. Turn the right-hand lobe of the liver forwards and note the corrugations on its dorsal surface. Cut through the epigastric vein at the point where it leaves the great omentum and turn it forwards. Investigate the origin of the duodenum from the right side of the gizzard. To the right of the duodenum lies the coiled intestine with which it is continuous.

There is no gall-bladder. There are two bile-ducts which may be seen by turning the liver forwards. They emerge from the liver close together. The left bile-duct leaves the liver between the right and left lobes and is the larger of the two. It passes backwards and opens into the proximal limb of the duodenum shortly after its origin from the gizzard. The right bile-duct leaves the right lobe of the liver and is smaller than the left. It passes backwards along the distal limb of the duodenum and opens into it after traversing about one-third of its length. There are three pancreatic ducts which vary slightly in position. Two ducts usually leave the middle of the pancreas and pass almost directly into the distal limb of the duodenum. The third duct leaves the middle, or anterior, part of the pancreas and runs forwards close to the right bile-duct and opens into the distal end of the duodenum. This opening may be seen by turning the duodenum over towards the left side in this region.

The hepatic portal vein may now be seen entering the liver just dorsal to the bile-ducts. It receives a large branch from the duodenal loop.

Cut through the walls of the air sacs on the left side and turn the gizzard forwards. On the left side dorsal to the liver may be seen the soft-walled proventriculus or anterior part of the stomach. The proventriculus passes into the thick-walled gizzard or posterior part of the stomach. The small intestine lies in coils on the right side, and in the middle line may be seen the straight rectum which is usually green or yellow in colour and bears a pair of rectal caeca about an inch before its opening into the cloaca.

The spleen lies in the mesentery between the proventriculus and the rectum. It is an oval red body usually about ½ inch long.

The hepatic portal system

Pick up the rectum with a pair of forceps and notice the large coccygeo-mesenteric vein lying in the mesentery just dorsal to the rectum.

The remaining branches of the hepatic portal vein may now be traced. Two gastric veins enter it from the proventriculus and gizzard just as it passes the gizzard. The duodenal branch has already been noted. It unites with a gastric vein and proceeds as the gastroduodenal vein. Lay the coiled small intestine out to the side without damaging its mesenteries and find the large mesenteric vein draining it and passing into the hepatic portal vein close to the spleen. The coccygeo-mesenteric vein enters the hepatic portal vein close to the spleen.

With the viscera laid out on one side as far as possible without cutting the mesenteries, observe the arrangement of the organs lying attached to the dorsal wall of the body cavity.

The kidneys and gonads

The kidneys are paired and each kidney is divided into three distinct lobes. They occupy the greater part of the roof of the body cavity. At the anterior end of the kidneys near the middle line may be seen the gonads, paired testes in the male or single left ovary in the female. In the female the left oviduct (there is no right) lies across the left kidney.

Veins in the sacral region
(Fig. 29)

The relations of the veins in this region to the body cavity should now be investigated. Trace the coccygeomesenteric vein back to its junction with the posterior mesenteric vein draining the posterior part of the rectum. Separate the muscles in the middle line just behind the posterior mesenteric vein and see the caudal vein emerging from the tail region. The caudal unites with the coccygeo-mesenteric and posterior mesenteric veins and immediately divides into a pair of hypogastric veins which run into the substance of the kidneys on either side. The hypogastric veins represent the renal portal veins of lower forms but are no longer portal in function. They run directly through the substance of the kidneys and emerge from their anterior ends between the gonads and unite to

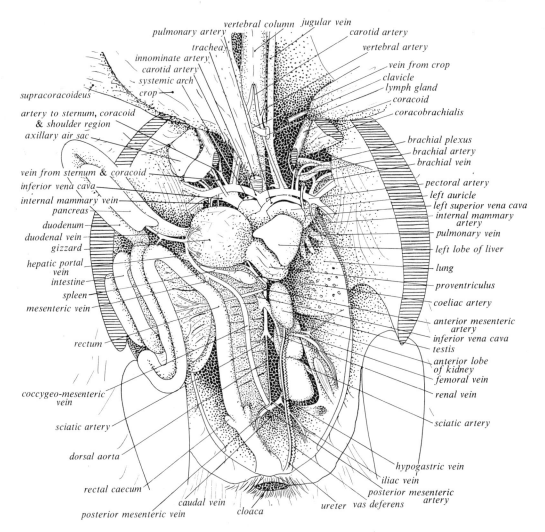

FIG. 29. Pigeon further dissected to show the arteries and veins around the heart and the contents of the body cavity. The coracobrachialis muscle has been laid back, the crop separated from its surrounding organs, and the trachea cut away. The viscera have been pulled over to the right and the heart is nearly covered by the liver and gizzard.

form the inferior vena cava. The inferior vena cava may be seen between and in front of the gonads. It passes forwards and enters the liver after a course of about $\frac{1}{2}$ inch. The inferior vena cava may be seen leaving the liver by turning the ventricle forward when this vessel will be seen leaving the liver on the right side and passing forwards into the right auricle.

The hypogastric vein in its course through the kidneys receives blood from the following veins. A pair of iliac (external iliac) veins enter the hypogastric veins on either side just as they enter the kidneys. The iliac veins drain the pelvic cavity. Entering the groove between the middle and posterior lobes of the kidney may be seen an artery and a vein lying side by side. These are the sciatic arteries and veins supplying the hind limb. The sciatic vein runs through the substance of the kidney to unite with the hypogastric vein. These vessels pass through the ilio-sciatic foramen of the pelvis. Between the anterior and middle lobes of the kidney the femoral vein may be seen returning from the limb and passing through the substance of the kidney to join the hypogastric vein. In the anterior part of the kidney may be seen a large vein lying over its surface. This is the renal vein which drains the kidney and unites with the hypogastric vein at the anterior end of the kidney.

In a fresh specimen owing to the bleeding that occurs it will be impossible to trace the junction of the veins lying in the substance of the kidney. If a preserved specimen is used for this purpose, the genital veins from the gonads, and in the female from the oviduct, may be seen entering the anterior part of the hypogastric veins.

Arteries of the body cavity and sacral region

Find the dorsal aorta lying in the middle line between the two kidneys. Trace it backwards where it runs ventral to the hypogastric vein in the middle line. It now passes backwards as the caudal artery which lies dorsal to the caudal vein.

Now proceed to trace the various branches from the dorsal aorta from behind forwards.

The posterior mesenteric artery leaves the dorsal aorta just before the latter becomes the caudal artery. The posterior mesenteric artery is very small and runs alongside the posterior mesenteric vein to supply the cloacal and posterior rectal regions.

The iliac arteries arise near the point where the dorsal aorta crosses the hypogastric vein. They run alongside the iliac veins and supply the pelvic region.

The sciatic arteries have already been noticed passing through the ilio-sciatic foramen along with the sciatic veins. Their origin from the dorsal aorta can be seen opposite the junction of the anterior and middle lobes of the kidney.

The reno-femoral arteries arise next and divide into renal arteries supplying the middle and posterior lobes of the kidney and the femoral artery proceeding to the limb. The femoral arteries may be seen lying alongside the femoral veins on either side by loosening the connective tissue along the outer anterior border of the kidneys. The origin of these arteries from the dorsal aorta lies above the base of the inferior vena cava, but this investigation should be postponed until after the examination of the urinogenital system.

The dorsal aorta now passes from view as it runs below (dorsal to) the union of the hypogastric veins. Here it gives off a pair of reno-lumbar arteries to the anterior lobe of the kidneys and the body wall.

Pull the viscera well to the right side in the anterior region of the body cavity. The anterior mesenteric artery will be seen rising from the mid-dorsal line just in front of the gonads and passing ventrally behind the spleen to supply the small intestine. The course of this vessel to the intestine may easily be traced.

Between the spleen and the proventriculus will be seen the coeliac artery running at first parallel with the anterior mesenteric artery. At the side of the proventriculus the coeliac artery may be seen to divide into arteries supplying the proventriculus, gizzard, small intestine, spleen, and liver.

Carefully follow the coeliac and anterior mesenteric arteries to their origin from the dorsal aorta which here lies between the lungs.

In the female the genital arteries are given off to the gonads and their ducts from the renolumbar and reno-femoral arteries.

Urinogenital system (Fig. 29)

The kidneys have already been described. The ureters are paired and arise from the posterior border of the anterior lobes of the kidneys. They run backwards along the ventral surface of the kidney and then straight back to open into the dorsal wall of the cloaca.

The adrenal bodies are small yellow structures near the anterior end of the kidneys and close to the hypogastric vein.

Male

The testes are paired oval bodies attached to the inner border of the anterior lobes of the kidneys. They vary very considerably in size; at the breeding season they are from $\frac{3}{4}$ to 1 inch in length while in the winter they shrink to about $\frac{1}{4}$ inch in length.

The vasa deferentia (mesonephric ducts) emerge from the inner side of the posterior end of the testes and run backwards along the outer border of the ureters as paired coiled tubes. The walls of the vasa deferentia are opaque while the walls of the ureters are thin and transparent. The vasa deferentia open into the cloaca on the dorsal wall beside the ureters. Just before their opening into the cloaca the vasa deferentia dilate to form the small vesiculae seminales.

Female

A single left ovary is present. In the embryo two ovaries are developed, but in the adult the left alone persists. The ovary is irregular in shape and ova in various stages of development can usually be seen projecting from its surface. The left oviduct is well developed and is the functional oviduct, while the right is a small short tube opening into the cloaca on the right-hand side.

The funnel of the left oviduct can be seen as a membranous structure extending from the anterior border of the ovary to the posterior border of the kidney. The walls of the oviduct are thickened as it runs backwards.

The egg when discharged from the ovary consists of the 'yolk' only. It is received by the funnel of the oviduct. On its passage down the oviduct to the cloaca it is fertilized and is surrounded by an albuminous substance, 'the white', while the shell is deposited over it by the hindmost portion of the duct. The oviduct opens into the dorsal wall of the cloaca on the left-hand side just outside the opening of the ureter. There is a corresponding but smaller opening to the right oviduct on the right side.

Slit open the cloaca and rectum by a median ventral incision. The rectum opens into the ventral wall of the cloaca.

The cloaca is peculiar in being indistinctly divided into three regions. The ventral anterior region into which the rectum opens is the coprodaeum. Behind this is the urodaeum into which the ureters and genital ducts lead. The urodaeum is separated from the most posterior chamber, the proctodaeum, by a small transverse partition from the dorsal wall. The proctodaeum opens to the exterior and in the young female receives the bursa Fabricii opening into its dorsal wall. This is a small tube only present in very young birds. It degenerates in the mature animal.

Vascular system

The arteries and veins of the body cavity and sacral region have already been examined.

Heart and great vessels (Fig. 29)

PRELIMINARY INSPECTION. Carefully separate the crop and oesophagus from the surrounding organs throughout the whole length of the neck so exposing the trachea and the vertebral column in this region. Carefully cut away the ventral portions of the clavicles and coracoids, taking care not to injure the blood-vessels in this region. Now remove the pericardial membrane from the heart and carefully dissect it away together with the connective tissue surrounding the vessels just in front of the heart. Find the systemic arch, a large whitish vessel, emerging from the heart between the two auricles. Close to its origin it gives off a pair of innominate arteries which proceed outwards. The main systemic artery, which is the single right systemic artery, can be seen proceeding directly downwards towards the dorsal side. It is continuous with the dorsal aorta which has already been seen in the body cavity. There is no left systemic arch.

INNOMINATE ARTERIES AND BRANCHES ARISING FROM THEM. The superior venae cavae may be seen lying parallel to and just behind the innominate arteries. In tracing the branches of these arteries be careful not to injure the venae cavae or the veins supplying them. Follow out the branches of the innominate artery on one side of the body, leaving the other side intact for the examination of the veins. About $\frac{1}{4}$ inch from its origin the innominate artery divides into subclavian and common carotid arteries. The large common carotid passes forwards along the neck, and after a course of about 2 inches penetrates the musculature below the vertebral column in the middle line. About $\frac{1}{2}$ inch before this, the vertebral artery leaves the common carotid, and passes directly dorsally to penetrate the vertebral column. The common carotid artery emerges from the neck musculature about $\frac{1}{2}$ inch behind the angle of the jaw. It then passes forwards and outwards to the base of the skull and divides into internal and external carotid arteries. In finding the external carotids be careful not to injure the veins in this region.[1]

From the subclavian artery just lateral to the origin of the carotid artery, there arises a small vessel from the anterior side of the innominate. This passes forwards and downwards towards the coracoid bone. Here it divides; one branch supplies the sternum and the ventral head of the coracoid, while the other branch passes dorsally to the shoulder region to supply the coracobrachialis muscle.

Immediately beyond this small vessel the subclavian artery divides into brachial and pectoral arteries. The brachial artery passes forwards and downwards (dorsally), penetrating through the brachial plexus and passing laterally to the base of the wing which it supplies. The pectoral artery divides beyond the origin of the brachial into two large pectoral arteries which pass outwards into the pectoralis muscle.

---

[1] In the bird, unlike the snake, the elongation of the neck causes elongation of the internal and external carotid arteries, the carotid arch remaining close to the systemic arch on either side. The external and internal carotid arteries then unite on either side anteriorly, and the external carotids disappear between this union and their origin from the systemic arches. Thus the single pair of carotids up the neck here represent the much elongated internal carotids (dorsal aortae) (see Fig. 24).

The subclavian artery of the bird is a 'ventral' subclavian not homologous with the 'dorsal' subclavian of the lizard and rabbit. It leaves the ventral part of the arterial system instead of the dorsal part.

These arteries have already been seen when the pectoral muscles were first laid back.

From the pectoral artery there arises on its posterior border a small internal mammary artery. This passes backwards along the inner surface of the ribs within the thorax.

There are two other minute arteries leaving the pectoral arteries which supply the musculature in this region.

PULMONARY ARTERIES. The pulmonary arch leaves the anterior left border of the right ventricle. In order to expose its base the connective tissue between the auricles must be very carefully removed. The pulmonary arch divides almost immediately into two pulmonary arteries. The left pulmonary artery passes outwards between the innominate artery and the left superior vena cava passing dorsal to the latter to enter the lung. The right pulmonary artery passes dorsal to the aortic arch and right superior vena cava to enter the right lung.

SUPERIOR VENAE CAVAE AND BRANCHES. The veins entering the superior venae cavae should now be examined on the other side of the body. Near its base the superior vena cava is formed by the union of a large jugular vein from the neck with the subclavian vein. Follow the jugular vein forwards up to the head. It lies close beneath the skin and unites with its fellow just behind the angle of the jaw. From its union the external and internal jugulars proceed forwards to the head. The branches from the jugular vein on its course along the neck are as follows. (1) At its end it receives a small vein draining the shoulder and passing along the anterior border of the lung. (2) As it passes forwards it receives on the median side several small vessels from the crop. (3) About 1 inch from the end of the jugular vein it receives a large branch from the crop. At the point of entrance of this vessel lies a small oval reddish body, the cervical lymph gland. (4) At about the same level the jugular vein receives on the lateral side a vein from the shoulder and on the dorsal side the (5) vertebral vein from the vertebral column. (6) Medially another branch is received from the crop and (7) laterally a large vein from the plexus of blood-vessels in the skin of the neck. (This plexus of blood-vessels is nearly always torn in removing the skin from the neck and a good deal of blood flows out from it.) (8) Anteriorly small veins are received from the oesophagus and trachea.

From the subclavian vein there arises ventrally a small vein which passes forwards, ventral to the subclavian artery, and drains the sternum and coracoid, lying parallel with the artery supplying these parts. The subclavian vein is formed by the union of the brachial and pectoral veins. The pectoral vein immediately divides again into two branches draining the pectoral muscles. These veins lie immediately dorsal to, and slightly on one side of, the corresponding pectoral arteries.

The brachial vein passes forwards and upwards, dorsal to the brachial plexus. Just above the brachial plexus the brachial vein draining the wing receives a branch on its anterior border from the shoulder region.

From the junction of the pectoral and subclavian veins the small internal mammary vein passes backwards close to the internal mammary artery draining the musculature between the ribs, the sternum, and the coracoid.

The superior vena cava on the other side is formed in a similar manner. Now trace the course of the superior and inferior venae cavae into the right auricle. There is no sinus venosus, the veins opening directly into the right auricle. Turn the ventricle forwards and observe the left superior vena cava running across the left auricle and reaching the right.

PULMONARY VEINS. The pulmonary veins may now be seen more clearly to the left-hand side. They originate from the ventral side of the left lung and pass dorsal to the left superior vena cava and into the left auricle. The union of the right and left pulmonary veins cannot be seen without disturbing the venae cavae.

CHAMBERS OF THE HEART. The heart of the pigeon is four-chambered; there are two auricles and two ventricles and the separation of the arterial from the venous blood is complete.

Without removing the heart cut open the right and left ventricles. Observe that the right ventricle has thin walls as compared with the very thick walls of the left ventricle. The right ventricle is smaller than the left. Insert a probe into the right ventricle and push it forwards so that it finds its way into the pulmonary arch. In the same way find the opening of the systemic arch out of the left ventricle.

Cut open the right and left auricles. Find the opening of the venae cavae into the right auricle. The opening of the venae cavae into the right auricle is guarded by a muscular fold forming the Eustachian valve. The large opening of the right auricle into the right ventricle lies in the posterior auricular wall. The right auriculo-ventricular valve is formed by one, and the left by two, muscular flaps.

The left auricle receives the opening of the pulmonary vein on its dorsal side.

Notice that the arrangement of the chambers of the heart and of the great vessels around the heart resembles the condition seen in the snake and the lizard, but the left systemic arch is here absent and the ventricular septum is fully formed.

The hind limb

Muscles of the pelvic girdle and hind limb
(Figs. 26(a), (b) 27)

Remove the skin from over the back, from the leg as far as the ankle, and from the ventral surface posteriorly as far as the anus. Note the fat deposited in the skin and the cutaneous blood-vessels. Remove the superficial subcutaneous fat by scraping the surface of the muscles with a scalpel. On the medial surface, note the small *ambiens* muscle beneath a fat deposit. It arises from the inner edge of the ilium and is inserted through the patellar ligament by a tendon which is connected with the flexors of the digits; it is a weak depressor of the femur and may help to flex the digits.

From the lateral aspect, dissect away the *tensor fasciae latae*, which is a thin sheet of muscle arising from fascia and from the posterior end of the iliac crest and inserted on the fascia over the thigh muscles. It is a weak levator of the femur. Observe and remove the *gluteus* which arises from the surface of the ilium and is inserted into the lateral face of the femoral trochanter. This muscle rotates the femur about its long axis and may also raise the femur.

Dissect the *extensor cruris* which consists of several parts: (1) the rectus femoris, arising on the anterodorsal end of the ilium, (2) the vastus, several parts arising from the anterior and lateral faces of the femur, and (3) the femorotibialis, a small muscle on the medial surface of the femur. These unite to form the patellar ligament which is inserted into the proximal end of the tibia; they are all extensors of the knee joint and the rectus femoris also protracts the femur. Pull on them and then cut them away. Observe the three short muscles, *iliacus, psoas*, and *iliotrochantericus*, arising from the ventral edge of the ilium and inserted on the head of the femur which they rotate and may protract.

Observe the large *biceps femoris*, arising from the iliac crest. The fibres of this muscle converge to form a tendon which passes through a tendinous loop and is inserted on the tibia between the heads of the gastrocnemius. Pull on the muscle and observe that it flexes the knee. Cut it away and remove the sciatic nerves and fat deposits that are thus exposed.

The *flexor cruris* consists of the following three parts: (1) lateralis, arising from the posterior edge of the ilium, (2) medialis, arising from the ischium, and (3) femoralis, arising from the distal end of the femur. The femoral part inserts on the lateral part and the lateral part then joins the tendon of the medial part and they insert together on fascia over the medial aspect of the tibia. The whole complex acts as a retractor of the femur and flexor of the knee. Cut away the flexor cruris and find the *caudofemoralis* which arises from beneath the tail and inserts on the proximal end of the femur; it receives an accessory head which arises from the iliac crest. This muscle may retract the femur but probably normally pulls the tail laterally. Remove it.

The *adductor femoris* consists of two parts, arising from the ventral border of the ischium and pubis and inserted on the distal end of the femur and proximal end of the tibia between the heads of the gastrocnemius. These retract the femur and may depress it. Cut away the adductors and observe the *quadratus femoris* which rotates and may retract the femur. This muscle arises from the lateral surface of the ilium and is inserted on the great trochanter. Remove the muscle and expose the *obturator externus*, arising on the pubis and inserted on the trochanter, and the tendon of *obturator internus* which arises on the medial surface of the ischium and forms a tendon passing through the obturator foramen to insert on the trochanter. These two muscles rotate the femur.

Observe the head of the femur and its manner of insertion into the acetabulum. Move the femur about and note the limitations on its movement in certain directions.

The muscles that are situated distally to the knee joint are very complicated. Bend and stretch the ankle joint and observe that as the ankle is flexed, the toes become flexed and when the ankle is extended, the toes also extend. When birds such as the pigeon alight on a horizontal bar the weight of the body causes the ankle joint to flex and the toes flex simultaneously, grasping the bar. This perching mechanism makes it possible for birds to sleep safely while roosting on the branches of trees. Muscles moving the toes arise from and are situated round the tibiofibula and they have long tendons which extend into the toes. The extensors (gastrocnemius) and flexors (tibialis and peroneus) of the ankle are well developed.

Nervous system and sense organs

The brain

The brain used for dissection should have been hardened in alcohol.

The brain lies very close to the roof of the skull and it will be necessary to use great care in removing the bone. It will be best to pick away in small pieces the roof and sides of the skull. Examine first the dorsal surface of the brain leaving it *in situ*.

DORSAL SURFACE. The olfactory lobes are very small and lie at the extreme anterior end of the brain just between the eyes.

The cerebral hemispheres are very large and relatively smooth.

In the division between the cerebral hemispheres at the posterior end is the small pineal body about the size of a pin's head. Immediately behind the cerebral hemispheres is the cerebellum. This is marked by a number of transverse fissures. Note that the optic lobes are not visible between the cerebral hemispheres and the cerebellum. They are just visible to one side of the cerebellum. On either side of the cerebellum and covered by the spongy bone of the auditory capsule is the floccular lobe of the cerebellum. This lobe is very difficult to dissect out and usually comes away with the removal of the bone.

The cerebellum extends backwards and covers the anterior portion of the medulla oblongata as far as the posterior end of the roof of the fourth ventricle. The medulla oblongata is thick and wide, merging rapidly into the spinal cord.

The brain of the pigeon is bent so that the axis of the brain is not a prolongation of the main axis of the spinal cord. The main point of flexure is in the region of the midbrain with the result that the posterior part of the brain lies at right angles to the anterior part. This flexure can be clearly seen in a dorsal view, but the point of flexure will not be determined until a section has been made.

Remove the brain from the skull by first of all exposing the small portion of the spinal cord near the medulla. Cut through the spinal cord near the medulla and gently raise the brain, cutting across the nerve roots as you come to them.

VENTRAL SURFACE. The olfactory lobes will be seen at the extreme anterior end. Behind these comes the ventral surface of the two hemispheres.

The optic chiasma is a large structure in the middle line. From the optic chiasma the optic tracts can be seen passing outwards running dorsally to the optic lobes. Behind the optic chiasma is a small depression into which projects the infundibulum (tuber cinereum). The pituitary body is usually torn from the infundibulum in removing the brain from the skull and will be found in a pit in the floor of the cranial cavity, the sella turcica.

Immediately behind the infundibulum is the large medulla. At the anterior border of the medulla will be seen a broad band of fibres forming the pons. Along the middle line of the medulla runs a shallow depression separating the pyramidal tracts.

The first ten pairs of cranial nerves will be found originating from the same region of the brain as in the skate. The nerves are for the most part rather small and difficult to locate. The hypoglossal nerve is intracranial in origin and arises from the medulla behind the vagus.

INTERNAL CAVITIES OF THE BRAIN. Make a vertical parasagittal section of the brain passing through the middle of one cerebral hemisphere. Note that the posterior portion of the cerebral hemisphere overlaps the optic lobes dorsally. The ventricle of the cerebral hemisphere is a very narrow cavity. It lies in the dorsal portion and has a thin dorsal roof. The great bulk of the hemisphere is constituted by the corpus striatum which bulges up from the floor of the ventricle.

Now make a median vertical sagittal cut passing between the division of the cerebral hemispheres. Note that the mid-brain dorsally is completely covered by a posterior extension of the cerebral hemisphere. Between the cerebellum and the cerebral hemisphere note the pineal body arising from the midbrain by a long stalk. Between the roof of the midbrain and the optic chiasma lies the cavity of the third ventricle, which may be obscured if the brain substance has been swollen by the preservative used. Note the thick floor and the thin roof of the medulla. The cerebellum is very large. The arrangement of the white and the grey matter in the cerebellum results in the formation of a tree-like structure of white matter, the arbor vitae. There is no cerebellar ventricle.

The flexures of the brain axis are now clearly seen. The primary flexure lies in the midbrain, bringing the posterior part nearly at right angles to the anterior part. A second nuchal flexure lies in the medulla at an angle to the spinal cord. A third pontal flexure lies in the region of the pons below the cerebellum. The brain is here bent ventrally in the opposite direction from the other two flexures.

The spinal cord

Remove the feathers and skin from the body along the region of the spinal cord.

Pare away the neural arches so as to expose the spinal cord. Note that in the region of the brachial plexus the spinal cord is enlarged.

In the lumbar region there is another enlargement of the spinal cord. This lumbar enlargement has a thin roof and the walls of the dorsal part of the spinal cord have diverged to produce an elongated cavity, the sinus rhomboidalis. Behind the lumbar enlargement the spinal cord rapidly narrows in the caudal region.

The eye

Remove the eyeball from its socket by cutting through the eye-muscles and the optic nerve. The relationships of the different layers will be found to be essentially the same as in the skate.

Place the eyeball in a dish of water and with a sharp scalpel or razor make a sagittal cut.

Note the curved cornea on the outside. The iris is yellow. Outside the iris is a hard ring of bony sclerotic plates arranged round the eye.

Inside note the lens. Outside the lens and below the sclerotic plates is the choroid coat. The retina is black and very soft. The entry of the optic nerve is seen as a yellow spot at the back of the eye and close beside it is a structure peculiar to birds known as the pecten. This is a black folded structure projecting into the posterior chamber of the eye and attached near the entrance of the optic nerve. It is formed of a fold of the choroid projecting through a slit in the retina.

A LIST OF THE NAMES FOR MUSCLES USED IN THIS MANUAL WHICH DIFFER FROM THOSE USED BY FISHER[1] (1946) AND CHAMBERLAIN[2] (1943)	PRESENT NAME	FISHER'S NAME	CHAMBERLAIN'S NAME
	Adductor femoris	Adductor profundus superficialis	Adductor
	Biceps femoris	Iliofibularis	Biceps femoris
	Caudofemoralis	Caudofemoralis	Cruratis caudalis
	Coracobrachialis	Coracobrachialis posterior	Coracobrachialis ventralis
	Coracohumeralis	Coracobrachialis anterior	Coracobrachialis dorsalis
	Cutaneus	Serratus meta-patagialis	Metapatagialis
	Deltoideus	Deltoideus	—
	Tensor patagii	Tensor patagii brevis and longus	Patagialis brevis and longus
	Deltoideus major	Deltoideus major	Deltoideus
	Deltoideus minor	Deltoideus minor	Spinatus
	Extensor cruris	—	—
	Rectus femoris	Iliotibials anterior	Sartorius
	Vastus	Vastus medialis and lateralis	Quadriceps femoris
	Femorotibialis	Femoritibialis	—
	Flexor cruris	Flexor cruris medius and lateralis	Gracilis, semimem-branosus and semitendinosus
	Gluteus	Gluteus profundus and piriformis	Gluteus
	Pectoralis	Pectoralis super-ficialis	Superficial pectoral
	Quadratus femoris	Ischiofemoralis	Quadratus femoris
	Rhomboideus	Rhomboideus	Trapezius
	Scapulohumeralis	Dorsalis scapulae	Teres major
	Supracoracoideus	Supracoracoideus	Deep pectoral (subclavis)
	Tensor fasciae latae	Iliotibialis lateralis	Tensor fasciae latae

[1] FISHER, H. I. (1946) Adaptations and comparative anatomy of the locomotor apparatus of New World vultures. *Am. Midl. Nat.* **35**, 545–727.
[2] CHAMBERLAIN (1943) *Atlas of avian anatomy* (Michigan, U.S.A.).

10 *The rat*

Rattus norvegicus

External features THE general shape is not unlike that of the lizard, but in life the legs are held beneath the body instead of laterally and the tail is more markedly differentiated from the body.

Except for the soles of the feet and the tip of the snout, the body is everywhere covered with hairs; these are modified epidermal structures which are diagnostic of mammals. On the head are groups of long hairs, the vibrissae, which are used as tactile sense organs and have sensory bulbs at their bases. The mystacial vibrissae, on either side of the nose, are the best developed; note also the superciliary (above the eyes), genal (between eye and ear), and mental (below the lower jaw) groups of vibrissae. The tail bears scales arranged in rings and has a sparse covering of hairs.

The head is connected with the trunk by a short, scarcely noticeable neck. The snout is pointed and bears two external nostrils separated by a naked patch of skin, the rhinarium. The upper lip is cleft, exposing the long chisel-like upper and lower incisors (one pair of each). Open the mouth and note the long diastema between the incisors and the molars. Flaps of skin extend inwards across the diastema and separate an anterior chamber from the rest of the mouth. Thus the rat can gnaw wood and other materials without particles entering the main part of the mouth. There is a fleshy tongue and the palate is strongly ridged transversely.

The eyes are situated laterally and near the top surface of the head; there are upper and lower eyelids and a nictitating membrane attached to the anterior edge. The ears each have a well-marked pinna which can be moved by groups of muscles inserted at the base.

The trunk is wider than the neck and the fore and hind limbs are well separated. There are four clawed digits on the fore foot (pollex reduced) and five clawed digits on the hind foot. The limbs are all short but the hind limbs are longer than the fore limbs. The elbow and knee joints are enclosed in the skin over the trunk.

The anus is situated at the base of the tail and in a mature male rat may be concealed by the large scrotal sacs. The penis lies anterior to the scrotum and may be erected by squeezing the base with forceps. In a female rat the genital and urinary apertures lie anterior to the anus and the urinary opening is raised on a small papilla, the clitoris. The nipples of the mammary glands (another diagnostic feature of mammals) form two irregular rows. There are usually six on each side of the body, the most anterior (thoracic) medial to the fore limb, then two close together behind the fore limb (axillary glands), one (abdominal) anterior to the hind leg, and two (inguinal) medial to the hind leg. Each nipple is a small raised structure, white in the non-nursing albino rat.

General instructions for dissection

Pin the rat on a board with its ventral surface uppermost. Lift the skin midway between mouth and anus and cut through it. Cut forwards in the mid-ventral line to the lower jaw and backwards to the anus cutting round the penis or openings of urethra and vagina. Loosen the skin, using fingers or scalpel, and pin it out so that the elbow and knee joints and outline of the trunk are exposed, taking care not to injure the blood vessels in the axilla and groin.

In the female rat, note the mammary glands, forming transparent masses of fatty tissue, and remove these attached to the skin. Note their blood supply in the neck, thoracic, axillary, and inguinal regions. Notice also the pair of preputial glands on either side of the urethral opening.

The abdomen
(Figs. 30, 31)

Lift the abdominal wall and cut forwards in the midline until the xiphoid cartilage (at the posterior end of the sternum) is exposed and backwards to the base of the penis or clitoris. Cut laterally from the xiphoid cartilage along the edges of the thorax and pin out the flaps of abdominal wall. Turn back the large sheets of fat usually present in the posterior part of the abdomen.

Without moving the viscera, identify the liver, stomach, spleen, left kidney, coils of intestine, large caecum, and, in the female, uterus.

Solar plexus
(Fig. 30(b))

Gently move the viscera to the animal's right side, exposing the left renal vein and the origins of the coeliac and anterior mesenteric arteries (these may be enclosed in fat). Pour on a small amount of 70 per cent alcohol to make nerves opaque. Very carefully dissect away fat and lymphoid tissue, exposing the aorta and the *solar plexus*. Identify the coeliac and anterior mesenteric sympathetic ganglia (close together and between the arteries of the same names), the cardiac sympathetic ganglion (by the aorta), and the greater and lesser splanchnic nerves (these arise from the ganglia of the sympathetic chains).

Alimentary canal

Separate the rectum from the duodenum and manipulate the alimentary canal looking for branches of the coeliac and anterior mesenteric arteries and of the hepatic portal vein. The *liver* has four lobes: the left is large and undivided, the median or cystic has a deep fissure, the right is large and partly divided, and the caudate or Spigelian lobe is wrapped round the *oesophagus*. The oesophagus enters the kidney-shaped *stomach* which is divided into two regions: the white part contains no digestive glands whereas the pink, highly vascular part secretes pepsin, acid, and rennin and leads through the pyloric sphincter to the *duodenum*. The *bile-duct* enters the first part of the duodenal loop (there is no gall-bladder but the release of bile into the intestine is controlled by a sphincter at the duodenal end of the duct). The *pancreas* is diffuse and pinkish in colour with many ducts entering the duodenum; it lies within the duodenal loop and also extends round the stomach to the *spleen*. The *ileum* (rest of the small intestine) is very much coiled so that its total length is about six times the length of the head and trunk of the animal. An ileocaecal valve is present at the junction of ileum, caecum, and colon and there are masses of lymphoid tissue. The blind end of the wide *caecum* forms a poorly defined *appendix*. The *colon* (large intestine) is short and the diagonal markings on its walls correspond to internal folds. The colon leads into the longer, narrower *rectum* which may contain faecal pellets and opens at the anus.

Blood vessels of the alimentary canal

The *coeliac artery* supplies the stomach (gastric artery), spleen (lienal artery), and liver (hepatic artery). The *anterior mesenteric artery* supplies the intestine and caecum (pancreatico-duodenal, many intestinal and three colic arteries) and the *posterior mesenteric artery* supplies the rectum.

Blood is drained from the gut and conveyed to the liver by factors of the *hepatic portal vein* which form three groups: lienal veins, draining the spleen, pancreas, and stomach; pyloric veins, draining the pancreas, stomach, and duodenum: superior mesenteric veins, draining the ileum, caecum, and colon. The veins from the intestine are usually very clear and easier to distinguish than the arterial branches.

Ligature the coeliac and anterior mesenteric arteries close to the aorta and the hepatic portal vein where it leaves the

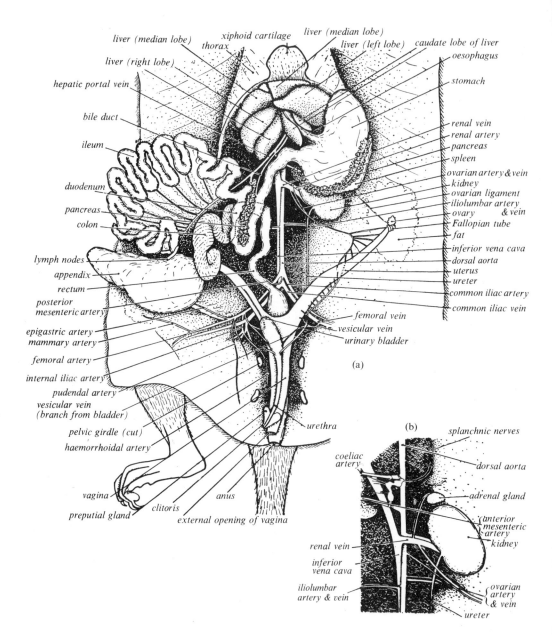

liver (median lobe)
xiphoid cartilage
liver (median lobe)
thorax
liver (left lobe)
caudate lobe of liver
liver (right lobe)
oesophagus
hepatic portal vein
stomach
bile duct
renal vein
renal artery
ileum
pancreas
spleen
ovarian artery & vein
duodenum
kidney
ovarian ligament
pancreas
iliolumbar artery
colon
ovary   & vein
Fallopian tube
fat
lymph nodes
inferior vena cava
appendix
dorsal aorta
rectum
uterus
posterior
ureter
mesenteric artery
common iliac artery
common iliac vein
epigastric artery
femoral vein
mammary artery
vesicular vein
femoral artery
urinary bladder
internal iliac artery
pudendal artery
vesicular vein
(branch from bladder)
pelvic girdle (cut)
urethra
haemorrhoidal artery
(a)
(b)
splanchnic nerves
coeliac
artery
dorsal aorta
adrenal gland
vagina
anus
anterior
mesenteric
preputial gland
clitoris
artery
external opening of vagina
kidney
renal vein
inferior
vena cava
ovarian
artery
iliolumbar
& vein
artery & vein
ureter

FIG. 30. Female Rat. (a) Abdominal and pelvic viscera; the xiphoid cartilage is pulled forwards, fat and part of the pancreas have been removed to display vessels, the arteries supplying the gut are not drawn, the ventral part of the pelvic girdle has been cut away, and branches of the common iliac artery are shown on the animal's right. Veins running parallel with the posterior arteries are frequently omitted. The left ovarian vein is drawn entirely, but the ovarian artery which runs with it is shown anteriorly only, and arises differently in the two figures.

(b) The alimentary canal and fat are removed to display the solar plexus and adjacent vessels.

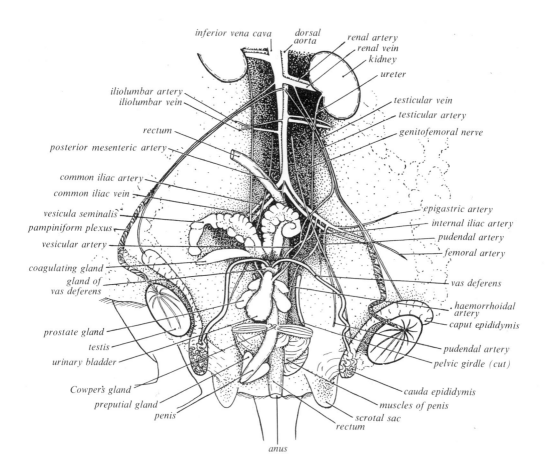

inferior vena cava

dorsal aorta

renal artery
renal vein
kidney

ureter

iliolumbar artery
iliolumbar vein

testicular vein
testicular artery
genitofemoral nerve

rectum

posterior mesenteric artery

common iliac artery
common iliac vein

epigastric artery
internal iliac artery
pudendal artery
femoral artery

vesicula seminalis
pampiniform plexus
vesicular artery

coagulating gland
gland of vas deferens

vas deferens

haemorrhoidal artery
caput epididymis

prostate gland
testis
urinary bladder

pudendal artery
pelvic girdle (cut)

Cowper's gland
preputial gland
penis

cauda epididymis
muscles of penis
scrotal sac
rectum

anus

FIG. 31. Male rat. Urinogenital system and blood vessels of the trunk and posterior region. The ventral part of the pelvic girdle has been cut away, and the branches of the common iliac artery are dissected out on the animal's left side only. Branches of the common iliac veins are not shown. On the right the coagulating gland has been separated from the vesicula seminalis.

duodenal loop and cut these blood vessels on the side of the gut. Remove the alimentary canal by cutting the oesophagus close to the stomach and the rectum about 1½ inches from the anus.

The male urinogenital system (Fig. 31)

Carefully separate the *kidneys* from the fat deposits round them and find the *ureters* leaving them. The right kidney is anterior to the left one and each ureter leaves the middle of the internal edge of the kidney. Anterior to each kidney is a brownish *adrenal gland*. The genitofemoral nerves, passing ventral to the lumbar muscles, should not be confused with the ureters.

Cut through the ventral wall of one *scrotal sac* and display its contents, carefully separating the large fat deposits that are usually present. Identify: *testis, caput and cauda epididymis, vas deferens* passing forwards from cauda epididymis. The *testicular artery*, leaving the aorta just behind the level of the left kidney, forms a network of capillaries (the pampiniform plexus) as it passes through the fat and then supplies the testis. On the left side only (usually) there is an anastomosis with the vesicular artery (see later).

Clear the muscles from the pubic symphysis, cut through the bones on either side of it and remove it. Carefully separate the structures in this region. Pull the *urinary bladder* posteriorly and find the ureters passing dorsal to the vasa deferentia and entering the neck of the bladder. The vasa deferentia pass through the *median gland of the vas deferens* and then enter the neck of the bladder. Associated with each vas deferens is a large, curved, sacculated *vesicula seminalis* and, in the same connective-tissue sheath, a large *coagulating gland*.

Turn the urinary bladder anteriorly and dissect away fat to expose the pair of *prostate glands*, each of which is partially subdivided. The bladder leads into the *urethra* which bears a pair of small *Cowper's glands* at the level of the posterior border of the pelvis. The urethra ends in the *penis*. Note the muscles and ligaments of the penis and find the pair of *preputial glands* at its tip.

Dissect away connective tissue and muscles behind the penis to expose the rectum leading to the anus.

The female urinogenital system
(Fig. 30(a))

Carefully turn the fat deposits posterior to the bladder forwards and expose the pubic symphysis. Cut through this lateral to the urethra on either side and remove the central part.

Carefully separate the *kidneys* from the fat deposits round them and find a *ureter* leaving the middle of the medial edge of each kidney. The right kidney lies anterior to the left one and anterior to each is a small brownish *adrenal gland*.

Follow the ureter through the fat deposits (where its path may be marked by a small blood vessel) towards the *urinary bladder*. Turn the bladder posteriorly and clear the fat behind it to show the ureter entering the neck after passing dorsal to the oviduct (uterus)—clear this carefully without injuring blood vessels. Follow the *urethra* to its opening in the clitoris between the pair of *preputial glands*.

Examine the reproductive system: on each side—small

pinkish *ovary*, connected by a ligament to lumbar muscles anterior to the kidney, a coiled *Fallopian tube* leading into the wider oviduct (*uterus*) which may contain developing embryos. The two uteri join to form the median *vagina* lying between the more ventral urethra and more dorsal rectum and leading to an external opening between the clitoris and anus.

Blood vessels of the abdomen (Figs. 30, 31)

The coeliac and anterior mesenteric arteries have already been ligatured and cut. Trace the *aorta* forwards; it emerges from the thorax immediately ventral to the dorsal muscles and on the left side. The right *renal artery* leaves the aorta between the coeliac and mesenteric arteries and, passing dorsal to the inferior vena cava, branches to give the right *adrenal artery* and then enters the right kidney dorsal to the right renal vein. The aorta usually passes dorsal to the left renal vein, giving off the left renal artery (which branches to supply the left adrenal gland) and then becomes ventral to the inferior vena cava.

The *inferior vena cava* receives the left and right *renal veins* (each of which receives an adrenal branch) and then, diverging towards the right side, passes forwards through the liver, receiving *hepatic veins*, and then through the diaphragm ventral to the oesophagus.

In the posterior part of the body the corresponding arteries and veins usually run very close together; the arteries are smaller, with thicker walls and appear pink in a fresh rat whereas the veins are blue.

The right *testicular artery* leaves the aorta opposite the left renal artery, passes ventral to the inferior vena cava and then, paralleled by the right *testicular vein*, through the pampiniform plexus to the right testis. The left testicular arteries and veins are variable; they may join the left renal vessels or arise directly from the aorta and vena cava posterior to the left renal vessels. Their distal course is similar to that of the right testicular vessels except that the left artery usually forms an anastomosis with the vesicular artery.

The origins of the *ovarian arteries and veins* are similar to those of the testicular vessels, being variable on the left side. These vessels supply the ovaries and Fallopian tubes and anastomose with the uterine arteries and veins (see later).

Posterior to the testicular or ovarian vessels are the *iliolumbar artery and vein* on each side, the left being anterior to the right. These supply the lumbar muscles and fat deposits.

The aorta then gives off the median *posterior mesenteric artery* to the rectum and divides into a pair of *common iliac arteries*. The inferior vena cava is formed by the fusion of a pair of *common iliac veins*. The common iliac vessels have two main branches: the *femoral artery and vein* supply the hind leg and can be traced through the abdominal muscles to the inner surface of the leg and the *internal iliac artery and vein* supply the dorsal muscles of the pelvis. The further branching of the blood-vessels is variable and may be asymmetrical but the following vessels can usually be identified by tracing their destinations: the *vesicular artery and vein* supply the bladder and genital organs near it and in males the left artery usually anastomoses with the left pampiniform plexus (of the testicular artery). The *haemorrhoidal artery and vein* supply the tissues round the anus. The *pudendal artery and vein* supply muscles of the pubic region. The *epigastric artery and vein* supply the posterior abdominal muscles. In males there is usually a vein from the pampiniform plexus on the left side.

In females, the *uterine artery and vein* (not labelled in Fig. 30) supply the uterus and anastomose on each side with the ovarian vessels. Branches from the femoral vessels usually supply the abdominal and inguinal mammary glands.

Raise the posterior ends of the aorta and inferior vena cava and find the *caudal artery and vein* in the midline between the body muscles; these supply the tail. Anterior to the caudal vessels find several median, dorsal *lumbar arteries and veins* supplying the vertebral muscles and vertebral column.

Muscles of the fore limb and pectoral girdle (Figs. 32, 33, 34)

It is convenient to examine these muscles now before proceeding with the rest of the general dissection. The student is advised to determine the actions of the muscles by pulling on each one as it is cleared from the surrounding connective tissue. Some of the muscles form thin sheets which are nearly transparent in the fresh animal and are therefore difficult to separate. These muscles are easier to distinguish in a rat preserved in formalin, but the preserved muscles tend to be brittle and it is not easy to determine their actions.

External features and skeleton

Observe the general form of the body with its slightly humped vertebral column. The legs are held below the body with the elbow directed backwards and the knee forwards. Move the limbs about and note the mobility in various planes of the shoulder and hip joints. If possible, observe a living rat and watch the way in which it moves and uses its limbs.

Examine the skeleton of a rat and note the following features that are characteristic of mammals: the pectoral girdle consisting of the clavicle and the scapula with its spine, acromion process, and coracoid process; the sternum consisting of sternebrae which articulate with the ventral ends of ribs; the pentadactyl fore limb including the ulna with an olecranon process; the pelvic girdle which articulates with the fused sacral vertebrae and has an enlarged obturator foramen and well-developed pubic symphysis; the pentadactyl hind limb; the vertebral column differentiated into cervical, thoracic, lumbar, sacral, and caudal regions.

The muscles of the pectoral girdle and fore limb

Skin one fore limb and the neck and chest of the rat beginning with a cut in the mid-ventral line and dissecting the skin away dorsally. Notice and cut through the cutaneous muscle that arises from the skin over the shoulder and is inserted near the head of the humerus.

From the dorso-lateral aspect, observe the following parts of the *trapezius*: (1) the *spino-trapezius*, arising from the spines of posterior thoracic and lumbar vertebrae and inserted on the dorsal third of the spine of the scapula, (2) the *acromio-trapezius*, arising from spines of cervical vertebrae and inserted on the acromion process and ventral two-thirds of the spine of the scapula, and (3) the *clavo-trapezius*, arising from the dorsal posterior surface of the skull and inserted on to the clavicle. These are all thin sheets of muscle partly covered with connective tissue; the acromio- and clavo-trapezius protract the scapula and clavicle respectively while the spino-trapezius retracts the scapula. Cut away the parts of the trapezius and remove the 'hibernating glands' that lie in the dorsal space between the two scapulae. Find the following thin sheets of muscle that may be difficult to separate from each other: the *occipito-scapularis*, which arises from the skull and is inserted on the dorsal edge of the scapula and its spine and protracts the scapula; the *rhomboideus*, which arises from cervical vertebrae and is inserted on the dorsal edge of the scapula which it protracts and elevates; the *levator scapulae*, which arises from cervical vertebrae and lies below the rhomboideus. Cut these muscles away.

Lay the rat on its back and remove the fatty tissue and lymph glands from the neck, taking care to avoid cutting the jugular vein. Observe the *cleidomastoideus* and *omotransversus* arising from the skull (mastoid process and basioccipital respectively) and inserted on the clavicle and the acromion process of the scapula respectively. These protract the clavicle and scapula. Remove them, exposing *omohyoideus* which arises from the hyoid plate and is inserted on the anterior edge of the scapula. Cut this away. All the muscles that have been observed so far, except the cutaneus, move the pectoral girdle relative to the body.

The *pectoralis* has several parts arising from the sternum and inserted on the humeral crest (deltoid ridge) and lesser tuberosity. These depress and retract the humerus. Part of the pectoralis also inserts on the coracoid process of the scapula.

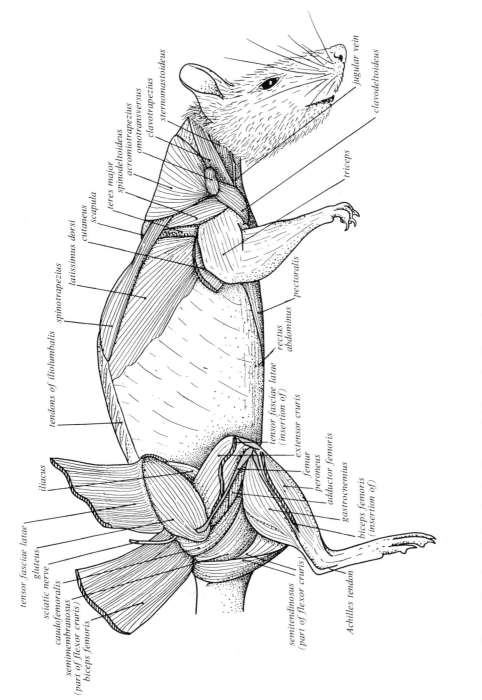

FIG. 32. Rat. Lateral view showing the superficial muscles. Cut surfaces are cross-hatched, tendons are white and bone exposed or nearly so is shown by mechanical tint. The skin, fat, connective tissue and lymph nodes have been removed. The cutaneus muscle has been cut away from the skin and biceps femoris and tensor fasciae latae have been reflected from their insertions on the fore leg and extensor cruris respectively.

jugular vein

clavodeltoideus

triceps

sternomastoideus

clavotrapezius

omotransversus

acromiotrapezius

spinodeltoideus

teres major

scapula

cutaneus

latissimus dorsi

spinotrapezius

tendons of iliolumbalis

pectoralis

rectus
abdominus

tensor fasciae latae
(insertion of)

extensor cruris

femur

peroneus

adductor femoris

gastrocnemius

biceps femoris
(insertion of)

Achilles tendon

semitendinosus
(part of flexor cruris)

iliacus

tensor fasciae latae

gluteus

sciatic nerve

caudofemoralis

semimembranosus
(part of flexor cruris)

biceps femoris

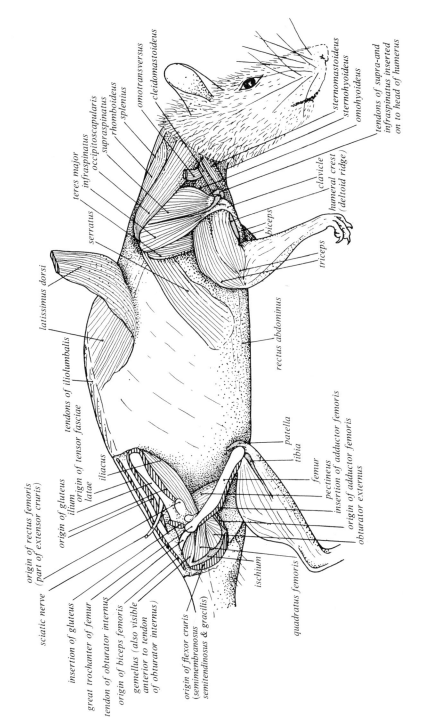

Fig. 33. Rat. Lateral view showing the deeper muscles and using the same conventions as in the preceding figure. The jugular vein and the following muscles have been removed: spino-, acromio- and clavo-trapezius, clavo- and spino-deltoideus, cutaneus, tensor fasciae latae, biceps femoris, gluteus, semimembranosus, semitendinosus, caudofemoralis, adductor femoris, and extensor cruris. Omotransversus and latissimus dorsi have been reflected from their insertions on the scapula and humerus respectively.

Labels (top/right side):
sternomastoideus
sternohyoideus
omohyoideus
tendons of supra- and infraspinatus inserted on to head of humerus
omotransversus
cleidomastoideus
splenius
rhomboideus
supraspinatus
occipitoscapularis
infraspinatus
teres major
serratus
latissimus dorsi
clavicle
humeral crest (deltoid ridge)
biceps
triceps
rectus abdominus

Labels (left/bottom side):
origin of rectus femoris (part of extensor cruris)
sciatic nerve
origin of gluteus
insertion of gluteus
ilium
tendons of iliolumbalis
origin of tensor fasciae latae
iliacus
great trochanter of femur
tendon of obturator internus
origin of biceps femoris
gemellus (also visible anterior to tendon of obturator internus)
origin of flexor cruris (semimembranosus semitendinosus & gracilis)
ischium
quadratus femoris
obturator externus
origin of adductor femoris
insertion of adductor femoris
pectineus
femur
tibia
patella

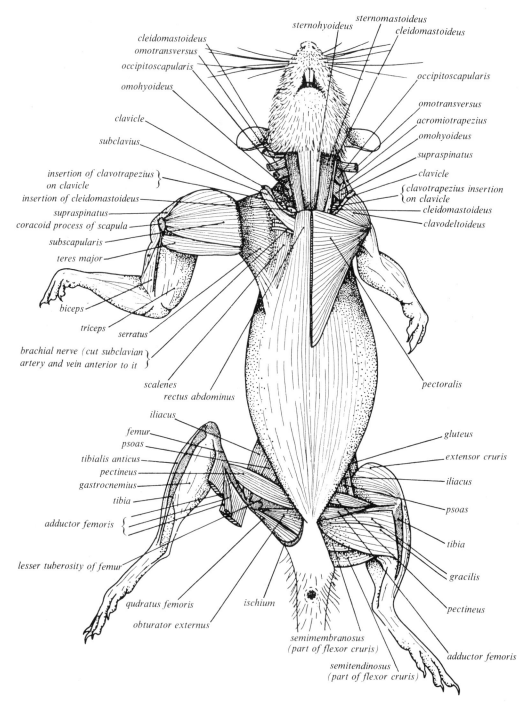

sternohyoideus
sternomastoideus
cleidomastoideus
cleidomastoideus
omotransversus
occipitoscapularis
occipitoscapularis
omohyoideus
omotransversus
acromiotrapezius
omohyoideus
supraspinatus
clavicle
subclavius
clavicle
clavotrapezius insertion on clavicle
insertion of clavotrapezius on clavicle
cleidomastoideus
insertion of cleidomastoideus
clavodeltoideus
supraspinatus
coracoid process of scapula
subscapularis
teres major
biceps
triceps
serratus
brachial nerve (cut subclavian artery and vein anterior to it)
scalenes
rectus abdominus
pectoralis
iliacus
femur
psoas
gluteus
tibialis anticus
extensor cruris
pectineus
iliacus
gastrocnemius
psoas
tibia
adductor femoris
tibia
lesser tuberosity of femur
gracilis
qudratus femoris
pectineus
obturator externus
ischium
adductor femoris
semimembranosus (part of flexor cruris)
semitendinosus (part of flexor cruris)

Remove the pectoralis and the fat deposits ventral and posterior to the arm. Note the *subclavius* connecting the first rib to the clavicle.

From the lateral aspect, observe the *latissimus dorsi*, which arises from the lumbodorsal fascia over the vertebral muscles and from the spines of thoracic vertebrae and forms a tendon which is inserted on the humeral crest. This muscle can retract and elevate the humerus and also rotates it. Cut it away.

Find the two parts of the *deltoideus*: acromio-deltoideus arises from the clavicle and acromion process of the scapula and spino-deltoideus arises from the scapular spine and from fascia posterior to it, both parts being inserted on the humeral crest (deltoid ridge) and on fascia. The acromio-deltoideus protracts the humerus and the spino-deltoideus elevates and retracts it. Remove both these muscles and separate the clavicle from the acromion process of the scapula. The limb and scapula can now be raised away from the body to which they are held only by the broad *serratus* muscle, which arises from the ribs and is inserted on the upper edge of the medial surface of the scapula. This muscle is a depressor of the scapula and is probably the principal muscle by means of which the body is supported from the pectoral girdle and fore limb. Cut it away from the edge of the scapula.

The scapula and limb can now be examined from their lateral and medial aspects. Note that almost the whole of the surface of the scapula is occupied by muscles which are inserted on the humerus. From the lateral aspect, note the large *supraspinatus*, arising from the surface of the scapula anterior to the spine and inserted tendinously on the head of the humerus which it protracts. Posterior to the spine is the large *infraspinatus*, also inserted tendinously on the head of the humerus, of which it is a rotator and possibly a protractor and levator. Infraspinatus and supraspinatus are believed to represent the supracoracoideus muscles of the reptiles. *Teres major* is probably a separated part of latissimus dorsi; it arises from the posterior border of the scapula and is inserted on the medial surface of the humerus which it retracts and rotates.

FIG. 34. Rat; ventral view showing superficial muscles on the animal's left and deeper muscles on the right. The skin, connective tissue, fat, and lymph nodes have been removed. On the left the clavotrapezius and jugular vein have been removed, and the omotransversus has been reflected from its insertion on the acromion process. On the right the following muscles have been reflected: omotransversus, occipitoscapularis, and omohyoideus from the scapula, cleidomastoideus from the clavicle, and adductor femoris from the pubis and ischium; the following have been removed: pectoralis, clavotrapezius, acromiotrapezius, deltoideus, extensor cruris, and flexor cruris group (semimembranosus, semitendinosus, and gracilis). The clavicle is separated from the scapula, and the fore limb is reflected outwards. The branchial plexus is cut, and in front of it lie the cut subclavian artery and vein.

Almost the whole of the medial surface of the scapula is occupied by *subscapularis* which is inserted on the lesser tuberosity of the humerus which it rotates and retracts.

Remove all these muscles and then observe the flexors (*biceps* group) and extensors (*triceps* group) of the elbow joint. Note that each of these has several parts, some arising from the humerus and some with tendinous origins from the scapula; the triceps inserts on the olecranon process of the ulna. In the more distal part of the limb are the extensors, flexors, and rotators of the wrist, and the muscles controlling the movements of the digits.

The thorax, neck, and mouth (Fig. 35)

It you have dissected the muscles fully, omit the next paragraph, which contains instructions for those who do not wish to study the muscular system.

Remove the fatty tissue and lymph glands from the neck, taking care not to injure the jugular vein and its branches; note the large parotid (salivary) gland behind the angle of the jaw. Carefully remove the *cleidomastoideus* and *omotransversus* muscles (see pp. 131, 134) and then remove the *pectoralis* muscle cutting it away from its origin on the sternum, exposing the clavicle; remove the subclavius muscle and the fat deposits ventral and posterior to the fore limb, thus exposing the nerves and blood vessels supplying the limb. Repeat on the other side of the rat.

Raise the xiphoid cartilage and cut through the ribs and intercostal muscles parallel to the edge of the thoracic cage. Tie a thread round the xiphoid cartilage and tie the other end round the base of the tail thus holding the diaphragm pulled backwards. Cut forwards on each side of the thorax, keeping the points of the scissors up and making the anterior ends of the cuts separate the sternum from the clavicles. Cut away the *sternohyoideus* muscle from the dorsal side of the sternum, remove the sternum with parts of ribs attached to it, and examine the blood supply to the intercostal muscles on their inner surface.

Examine the *diaphragm* and note the phrenic nerves supplying it, and the inferior vena cava, oesophagus, and aorta passing through it. Cut through the pleural connective tissue and examine the *lungs*. The left lung is single and the right is divided into four lobes. Push the left lung to the right and note the large *azygos vein* which drains blood from the intercostal veins and joins the left superior vena cava.

Carefully lift up the posterior end of the sternohyoideus muscle and turn it forwards, exposing the trachea. Identify the pink *thymus gland* covering the anterior end of the heart and carefully remove it without injuring the nerves and blood vessels below it. Cut away the thin *pericardium* and observe the two dark red *auricles* and large muscular ventricular region showing no obvious division into two. Carefully dissect the subclavian and jugular veins away from the clavicles, cut these bones near the joints with the scapulae and remove them.

The great veins

Turn the heart and left lung towards the right side and identify the branches of the *left superior vena cava*: the azygos vein joins it as it enters the right auricle; the *subclavian vein*, composed of brachial and axillary branches, joins the *jugular veins* ventral to the clavicle. The brachial vein is the main vessel draining the fore limb and the axillary vein drains the anterior mammary glands of the female. The internal jugular vein is small, lies close to the trachea and may branch. The *external jugular vein* originates from the junction of the anterior and posterior facial veins, which drain the structures inside and outside the skull respectively; it collects a cervical vein draining the neck and there are other small branches.

Move the left lung to the left and follow the left *phrenic nerve* beside the superior vena cava, between the lungs and to the diaphragm. Note the left bronchus entering the left lung and the left pulmonary artery and vein supplying it. Note the junction of the inferior with the left superior vena cava.

Move the heart over to the left side and follow the right phrenic nerve to the diaphragm. Note the junction of the *right superior vena cava* with the other two and the right pulmonary artery and vein supplying the right lung. The branches of the right superior vena cava differ from those of the left side only in the absence of the azygos vein which is present on the left side only and drains both sides of the thorax. (The right internal jugular vein is omitted from Fig. 35.)

The great arteries and the nerves near the heart

Turn the heart back to the median position and identify the *aorta* and *pulmonary artery*. Pour a little 70 per cent alcohol round the heart to make the nerves opaque.

Carefully remove connective and glandular tissue to expose the *left vagus nerve* on the inner side of the posterior end of the left superior vena cava. After crossing the aorta, this divides into cardiac depressor and visceral and recurrent laryngeal branches and the latter loops round the aorta and passes forwards alongside the trachea to the larynx. Clear the base of the *left carotid artery* taking care not to damage the vagus and sympathetic nerves on either side of it.

Clear the base of the *innominate artery* (not labelled in Fig. 35); it divides into the *right carotid* and *right subclavian arteries*. Find the *right vagus nerve* beside the innominate artery; the recurrent laryngeal branch of the vagus loops round the subclavian artery and then passes forwards along-

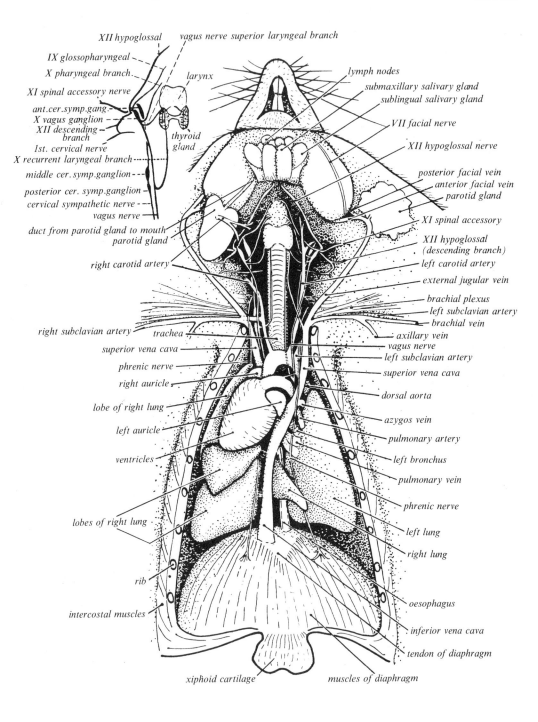

XII hypoglossal

vagus nerve superior laryngeal branch

IX glossopharyngeal

X pharyngeal branch

larynx

XI spinal accessory nerve

ant.cer.symp.gang.

X vagus ganglion

XII descending branch

thyroid gland

1st. cervical nerve

X recurrent laryngeal branch

middle cer. symp.ganglion

posterior cer. symp.ganglion

cervical sympathetic nerve

vagus nerve

duct from parotid gland to mouth

parotid gland

right carotid artery

right subclavian artery

trachea

superior vena cava

phrenic nerve

right auricle

lobe of right lung

left auricle

ventricles

lobes of right lung

rib

intercostal muscles

xiphoid cartilage

lymph nodes

submaxillary salivary gland

sublingual salivary gland

VII facial nerve

XII hypoglossal nerve

posterior facial vein

anterior facial vein

parotid gland

XI spinal accessory

XII hypoglossal (descending branch)

left carotid artery

external jugular vein

brachial plexus

left subclavian artery

brachial vein

axillary vein

vagus nerve

left subclavian artery

superior vena cava

dorsal aorta

azygos vein

pulmonary artery

left bronchus

pulmonary vein

phrenic nerve

left lung

right lung

oesophagus

inferior vena cava

tendon of diaphragm

muscles of diaphragm

side the trachea. Ligature the right superior vena cava at the level of the junction of the innominate artery with the aorta, cut through the vein and turn the cut end forwards without damaging any nerves or arteries. Follow the right subclavian artery to the fore limb and note the *brachial plexus nerves* supplying the limb.

Turn the heart to the right and ligature the left superior vena cava and turn the cut end forwards. Expose the origin of the *left subclavian artery* and follow it into the fore limb with the nerves of the brachial plexus. Move the left lung towards the right and follow the *aorta* which passes dorsal to the azygos vein and backwards to pass through the diaphragm.

Ligature the azygos vein and cut it close to the superior vena cava. Trace the *left cardiac depressor nerve* to the heart. Carefully dissect out the *left pulmonary artery* and note the narrow band of tissue, the *ductus arteriosus* (not labelled in Fig. 35), connecting it with the aorta.

Turn the heart to the left and trace the *right pulmonary artery* and the *right cardiac depressor nerve*.

The pulmonary veins can best be seen entering the heart when this is removed as described later.

The structures in the neck

The blood vessels and nerves at the base of the neck have already been described and so have the branches of the jugular veins.

Expose the *trachea* and *larynx* fully by removing the sternohyoideus muscle and identify the *thyroid gland*. Pour some 70 per cent alcohol over the neck to make the nerves opaque.

Lift up the posterior edge of the hyoid cartilage and remove it by cutting through the muscles attached to its dorsal surface. This will expose the large hypoglossal nerve (*cranial XII*) which supplies the muscles of the tongue and certain hyoid muscles; trace this nerve back to its origin from the condylar foramen and identify and clear the descending branch of the hypoglossal which supplies some neck muscles and forms an anastomosis with the *first cervical nerve*.

Cut away the stylohyoid muscle from the skull (styloid process) exposing the nerves leaving the foramen lacerum posterius. Note the vagus ganglion immediately posterior to the origin of this nerve from the foramen. The vagus (*cranial*

FIG. 35. Thorax and neck of Rat. The xiphoid cartilage is pulled backwards, the ribs and intercostal muscles are cut and the sternum removed, the thymus gland and pericardium are dissected away, the clavicles are removed, the parotid gland is turned outwards on the animal's neck, the right carotid artery is cut and its anterior end reflected forwards, the posterior part of the left cervical sympathetic nerve has not been drawn, and the right internal jugular vein has been omitted.

Inset. A diagram identifying the nerves in the neck shown on the animal's right side, and showing the position of the thyroid gland.

X) has the following branches: superior laryngeal, leaving the main nerve at the level of the thyroid and passing dorsal to the carotid artery to supply the larynx; pharyngeal, passing forwards to anastomose with the glossopharyngeal nerve; recurrent laryngeal, leaving the main nerve at the level of the subclavian artery on the right or aorta on the left side and looping dorsally to run forwards alongside the trachea to the larynx; cardiac depressor, supplying the heart; visceral, supplying the gut and other viscera.

The glossopharyngeal (*cranial IX*) nerve also leaves the foramen lacerum posterius; it passes forwards supplying some muscles and sense organs of the tongue. The spinal accessory (*cranial XI*) nerve, leaving the same foramen, supplies some of the muscles of the neck.

Carefully separate the carotid artery half way along the neck from the vagus, lying lateral to it, and the *cervical sympathetic nerve*, lying medial to it; cut the artery and turn it forwards. Follow the sympathetic nerve forwards and find the anterior cervical sympathetic ganglion slightly anterior and medial to the vagus ganglion. Anterior to this, the sympathetic nerve enters the middle ear through the carotid foramen. Follow the sympathetic nerve backwards and find the middle and posterior cervical sympathetic ganglia at the level of the brachial plexus. Try to follow the sympathetic trunk back through the thorax and abdomen where there are segmental ganglia and numerous branches to the viscera.

Note the *cervical spinal nerves*, supplying neck muscles, and find the origin of the *phrenic nerve*; this is usually from the fifth cervical nerve which is also the most anterior nerve contributing to the *brachial plexus*, which supplies the fore limb.

The heart and lungs

Cut through the trachea immediately posterior to the larynx and through the oesophagus dorsal to it. Raise the trachea, with oesophagus attached to it, and pull them backwards, cutting the connective tissue. Cut the innominate, left carotid and left subclavian arteries, and the aorta and remove the trachea with heart, lungs, and associated structures.

Remove the oesophagus from the isolated preparation and examine the rest from the dorsal aspect to see the trachea dividing into two *bronchi* each of which supplies one lung. Dissect out the *pulmonary arteries and veins* and cut them close to the lungs. Remove the lungs.

Examine the isolated *heart*. The two auricles are partly separated but the ventricles are firmly bound together. Cut through the ventricles transversely and note the thick muscular walls and the cavities of the two ventricles. Push a blunt seeker through the central cavity (left ventricle) and note that it emerges from the aorta. Push the seeker through the lateral cavity (right ventricle) and manœuvre it through the pulmonary artery. The seeker can also be pushed from each ventricle into the corresponding auricle.

Examine the outside of the heart and note the *coronary veins and arteries*, the latter arising from the base of the aorta and disappearing into the heart muscle.

The mouth

Cut through the cheek muscles (masseters, closing the mouth) on either side and force the mouth open firmly, dislocating the jaw articulations. Identify the upper and lower molar teeth of which there are three on each side and note the transverse ridges on the *hard palate*. Examine the *tongue*. On one side, cut along the side of the pharynx to the beginning of the oesophagus and reflect the lower jaw. Examine the *epiglottis* and push a seeker through the glottis into the trachea. Observe the single opening of the *internal nares* almost opposite the glottis and insert a seeker into the nasopharynx, dorsal to the *soft palate*. Insert a seeker through one nostril and manipulate it until it can be seen above the soft palate. Cut forwards along the midline of the soft palate from the internal nares to the posterior edge of the hard palate. Move a seeker along the side of the nasal passage starting opposite the edge of the hard palate and pressing on the pterygoid flange. Immediately posterior to the pterygoid, the point of the seeker can be inserted into the *Eustachian tube* which is connected with the middle ear.

Separate the lower incisors with a scalpel and note that there is no bony symphysis between the lower jaws.

The muscles of the pelvic girdle and hind limb
(Figs. 32, 33, 34)

Remove the skin from over the posterior half of the body and from one hind limb, exposing the base of the tail and the anterior end of the ilium. Slips of *cutaneus* muscles arise from the fascia over the limb and insert in the skin. Dorsally, note the strong tendons of the *iliolumbalis*, the chief extensor of the vertebral column, and ventrally, note *rectus abdominis*, which is a flat, thin band on the ventral side of the abdomen. Remove the connective tissue and deposits of fat from the surface and proximal end of the limb.

From the dorso-lateral aspect, dissect *biceps femoris*, arising from the spines of anterior caudal vertebrae and inserted as a flat band along the proximal two-thirds of the tibia and distal end of the femur. Note that this muscle is a flexor of the knee and a levator of the femur, then remove it carefully, exposing the sciatic nerve and fat deposits. Anterior to the biceps femoris is the thin sheet of *tensor fasciae latae*, arising from fascia over the vertebral column and inserted into fascia over the extensor cruris. Cut it away, exposing the mass of *gluteus* muscles which are very difficult to separate from each other. They arise from the anterior end, dorsal border, and lateral surface of the ilium and from fascia over the sacrum and insert fleshily and by tendons on to the great and third trochanters of the femur of which they are rotators and retractors. Remove the gluteus mass and dissect the *extensor cruris*, the extensor of the knee joint, a mass of muscle along the anterior border of the femur. Note that part of it, the rectus femoris, arises by a strong tendon from a process on the ilium and can protract the femur, while the rest of it, the vastus, arises from the proximal end of the femur. They insert together through the patellar ligament on to the tibia. Cut away the extensor cruris.

Examine next the *flexor cruris* muscles (semitendinosus, semimembranosus, and gracilis), which lie posterior to the femur; they retract it and flex the knee. Cut away the sciatic nerve and remove the fat deposits and dissect out the semi-tendinosus, which arises from caudal vertebrae immediately posterior to biceps femoris and from the posterior edge of the ischium and is inserted on the tibia. Remove the semitendi-nosus and observe the semimembranosus, arising from the posterior border of the ischium and inserted on the tibia. Anterior to this is the *caudofemoralis*, which arises from caudal vertebrae and is inserted on the distal end of the femur. This retracts and rotates the femur. Turn the rat on to its back, expose the medial surface of the leg, and dissect the two gracilis muscles. One is closely attached to the semimem-branosus and arises from the ischium, while the other arises from the pubic symphysis and is not easy to separate from the adductor muscles. Both parts insert on the tibia. Remove the flexor cruris and caudofemoralis.

The femur is depressed and may also be retracted by the *adductor femoris* group which is now exposed and consists of three parts, all arising from the pubis, pubic symphysis, and ischium and inserted separately on to the distal end of the femur, the shaft of the femur, and the third trochanter and its shelf. These may have been cut while dissecting the urino-genital system (p. 128). Remove the adductors, cutting through the nerves that may be seen to supply them.

Turn the femur backwards, cut away the femoral nerve and vein and observe the *pectineus* arising on the pubis and in-serted on to the proximal end of the femur, which it protracts. Lateral to the pectineus are the *psoas*, with a strong tendon, and the *iliacus*, both arising from the ventral surfaces of lum-bar vertebrae and both inserted on the lesser trochanter of the femur. They protract the femur and rotate it about its long axis. Remove the three muscles.

Turn the femur forwards and observe the following short

muscles: (1) the *quadratus femoris*, arising from the posterior border of the ischium and inserted on the lesser trochanter of the femur which it retracts—remove it; (2) the *obturator externus*, a depressor of the femur, arising from around the obturator foramen and inserted into a fossa below the trochanter; (3) the *gemellus*, arising from the upper border and posterodorsal margin of the ischium and inserted in the trochanteric fossa; (4) over the surface of gemellus passes the glistening tendon of the *obturator internus*, which arises on the medial surface of the ischium and inserts in the trochanteric fossa. Gemellus and obturator internus are rotators of the femur and are the antagonists of psoas and iliacus.

Examine the muscles of the lower leg and foot and note that those on the antero-medial aspect are flexors of the ankle and extensors of the digits, while those on the postero-lateral aspect are extensors of the ankle and flexors of the digits. The principal extensor of the ankle is the *gastrocnemius* with its Achilles tendon passing over the heel.

A LIST OF	PRESENT NAME	GREENE'S NAME
THE NAMES	Adductor femoris	Adductor brevis, longus, and magnus
FOR MUSCLES	Extensor cruris	Extensor quadriceps femoris
USED IN	Rectus femoris	Rectus femoris
THIS MANUAL	Vastus	Vastus externus, internus, and intermedius
WHICH DIFFER		medius
FROM THOSE	Gluteus	Gluteus medius and minimus and piriformis
USED BY	Omotransversus	Levator claviculae
GREENE[1] (1935)	Psoas	Psoas major
	Serratus	Serratus anterior
	Tensor fasciae latae	Tensor fasciae latae and gluteus maximus

[1] GREENE, E. C. (1935) Anatomy of the rat. *Trans. Am. phil. Soc.* **27** (n.s), 1–370.

11 *The brain of sheep*[1]

The membranes and blood-vessels

THE brain and spinal cord are closely invested with connective-tissue membranes or meninges. The outer of these, the dura mater, is a dense fibrous membrane closely adherent to the skull and vertebral column, and serves as nourishing membrane, or periosteum, to the bone. A fold of dura mater extends downwards between the cerebral hemispheres and between the cerebellum and cerebral hemispheres.

The pia mater lies within, and is very vascular and follows the contour of the brain very closely. In it the greater part of the blood-vessels of the brain lie. The pia mater is folded and highly vascular over the roof of the ventricles of the brain forming the choroid plexus. The blood supply to the brain is derived from the paired vertebral and internal carotid arteries.

The spaces within the brain and between the meninges and the bone are filled with cerebrospinal fluid.

In the hardened brain of the sheep parts of the brain membranes will already have been removed. Notice the blood-vessels in the pia mater supplying the brain. Below the midbrain lies a vascular ring, the circle of Willis. This vessel is supplied laterally by the internal carotids and posteriorly by the basilar artery, lying across the pons, and formed by the union of the two vertebral arteries. From the circle of Willis

---

[1] The brain of the sheep is best removed from the fresh skull and preserved and stored in the following fluid:

Chrome alum	2·5 g,
Copper acetate	5·0 g,
Glacial acetic acid	5·0 cm³,
Formol, 4 per cent	10·0 cm³,
Water	77·5 cm³.

Boil the chrome alum in water, when dissolved remove from the flame, add the copper acetate in fine powder and the acetic acid, stirring well. Let cool and add the formol. A fortnight's hardening is sufficient for practical purposes.

branches pass out to supply the internal ganglia of the thalamus and corpora striata and to supply the outer cortical portions of the brain by way of the pia mater.

Dorsal aspect of the brain

At the anterior end are the olfactory lobes, relatively small rounded masses, into whose anterior surfaces the fibres of the olfactory nerves enter. Posterior to them are the cerebral hemispheres. Their surfaces are much convoluted (cp. rabbit) giving a great surface area. Posteriorly the hemispheres cover completely the midbrain and reach the cerebellum (cp. lower types). The median vermis of the cerebellum corresponds to the entire cerebellum of lower vertebrates. Situated laterally are a pair of cerebellar hemispheres from which the floccular lobes arise. The vermis entirely covers over the fourth ventricle.

On one side cut away the posterior part of the cerebral

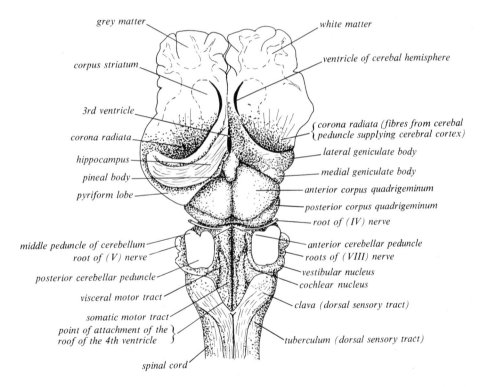

grey matter
white matter
corpus striatum
ventricle of cerebal hemisphere
3rd ventricle
corona radiata (fibres from cerebal peduncle supplying cerebral cortex)
corona radiata
lateral geniculate body
hippocampus
medial geniculate body
pineal body
anterior corpus quadrigeminum
pyriform lobe
posterior corpus quadrigeminum
root of (IV) nerve
middle peduncle of cerebellum
anterior cerebellar peduncle
root of (V) nerve
roots of (VIII) nerve
posterior cerebellar peduncle
vestibular nucleus
cochlear nucleus
visceral motor tract
clava (dorsal sensory tract)
somatic motor tract
point of attachment of the roof of the 4th ventricle
tuberculum (dorsal sensory tract)
spinal cord

FIG. 36. Sheep's brain partially dissected. The cerebral cortex has been cut down on the left side to expose the hippocampus, and on the right side the hippocampus has been removed exposing the midbrain fully. The cerebral peduncle and corona radiata have been cut into on either side. The cerebellum has been cut off exposing the fourth ventricle, the roof of which has also been removed.

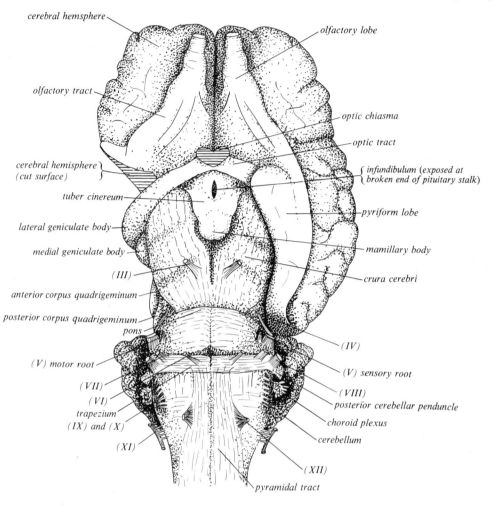

cerebral hemsphere

olfactory lobe

olfactory tract

optic chiasma

optic tract

cerebral hemisphere }
(cut surface) }

{ infundibulum (exposed at
{ broken end of pituitary stalk)

tuber cinereum

pyriform lobe

lateral geniculate body

medial geniculate body

mamillary body

(III)

crura cerebri

anterior corpus quadrigeminum

posterior corpus quadrigeminum

pons

(IV)

(V) motor root

(V) sensory root

(VII)

(VI)

(VIII)

trapezium

posterior cerebellar penduncle

(IX) and (X)

choroid plexus

(XI)

cerebellum

(XII)

pyramidal tract

Fig. 37. Ventral view of the brain of the sheep with the posterior part of the right cerebral hemisphere cut away. The cut surfaces of the cerebral hemisphere and of the optic chiasma are cross-hatched. The pituitary body has been removed.

hemisphere through the pyriform lobe so as to expose the midbrain. In the mid-dorsal line will be seen the narrow roof of the third ventricle and the large pineal body behind it. The large anterior corpora quadrigemina, or optic lobes, cover over dorsally the smaller, posterior corpora quadrigemina. Laterally and in front of the corpora quadrigemina are two swellings on either side. The anterior of these is the lateral geniculate body lying close to the cerebral hemisphere and the posterior, and more ventral, swelling is the medial geniculate body. The optic tract passes into the lateral geniculate body on either side. In front and within the lateral geniculate body lies the large cerebral peduncle which connects the cerebral hemispheres with the midbrain. The peduncle can only be seen after further dissection.

Ventral aspect of the brain (Fig. 37)

Carefully remove the brain membranes where necessary in order to expose the tracts, being careful not to pull off the nerve-roots.

From the olfactory lobes at the anterior end the olfactory tracts run backwards and outwards to terminate in the pyriform lobe, which forms the postero-ventral part of the cerebral hemispheres. Between the pyriform lobes lies the midbrain. At the anterior end is the optic chiasma and the optic tracts may be seen passing out laterally from it. Behind the optic chiasma is the elevated tuber cinereum with the mamillary body behind, but not sharply marked off from it. From the tuber cinereum hangs the pituitary body. The pituitary body may have been left behind in the sella turcica in the floor of the cranium, in which case its broken stalk will then be seen attached to the tuber cinereum and the infundibular cavity may be exposed.

On the sides of the midbrain will be seen bands of fibres, the crura cerebri, which connect the cerebral hemispheres with the medulla. The III nerve arises from the ventral posterior part of the midbrain, passing through the crura cerebri.

The pons is a heavy band of fibres crossing the anterior ventral part of the hind brain. Laterally the pons narrows and curls upwards to form the middle peduncle of the cerebellum. Between the middle peduncle of the cerebellum and the midbrain may be seen the root of the IV nerve which arises from the dorsal side of the brain between the cerebellum and the optic lobes and then passes ventrally. Immediately behind the middle peduncle of the cerebellum is the root of the V nerve. Its dorsal sensory root is large and the ventral motor root is small.

Behind the pons lies the transverse trapezium; some of its fibres turn forwards and run dorsal to the pons. The VII nerve arises behind the V and emerges through the trapezium. The VIII nerve and the trapezium originate from the lateral area acoustica. The area acoustica on either side consists of the restiform body (or posterior cerebellar peduncle) and the vestibular and cochlear nuclei. The restiform body connects the medulla with the cerebellum and lies close to the trapezium. The vestibular and cochlear nuclei consist of neurons forming bulges on the upper lateral walls of the fourth ventricle on either side.

From the posterior margin of the pons on either side of the ventral fissure run the longitudinal pyramidal or somatic motor tracts. Where the pyramids emerge lie the roots of the VI nerves on either side, and behind the restiform body lie the small numerous roots of the IX, X, and XI nerves. The XII nerve arises by many roots from the lateral border of the pyramids behind the level of origin of the XI nerve.

Sagittal section of the brain (Fig. 38)

The cortex of the cerebral hemispheres consisting of neurons can be seen lying outside the white matter formed by nerve fibres. The corpus callosum that connects both hemispheres is cut and lies obliquely across the middle of the cut surface of the cerebral hemisphere. From the middle of the corpus callosum a band of fibres, the fornix, curves ventrally.

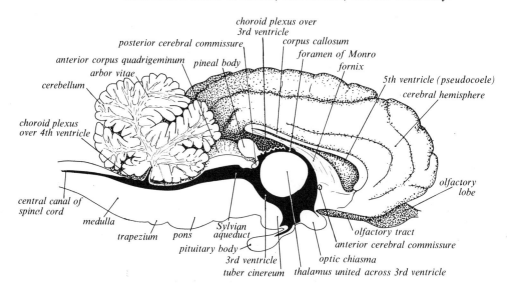

FIG. 38. Longitudinal half of the brain of a sheep viewed from the median side. The cavities within the brain are black.

The fornix forms part of the anterior boundary of the third ventricle, the cavity of which extends into the tuber cinereum and pituitary body. The third ventricle may be very small in size owing to swelling of the brain substance in preservation and hardening. The middle of the third ventricle is occupied by a round mass of tissue formed by the union of the lateral parts of the thalamus across the ventricle. Over the thin roof of the third ventricle lies the choroid plexus and behind it the pineal body. The brain cavity narrows below the corpora quadrigemina forming the Sylvian aqueduct connecting the third and fourth ventricles. The fourth ventricle narrows again posteriorly to the central cavity of the spinal cord.

In the ventral anterior wall of the third ventricle lies the anterior commissure connecting both sides of the brain. The posterior commissure lies in the roof of the midbrain just behind the base of the pineal stalk.

The cerebellum shows tree-like branching of the white matter within the grey matter, the arbor vitae. The cerebellar cortex is much folded and divided by sulci.

Completion of examination from dissections

In brains that have been further dissected the following additional structures should be noted.

In a brain with the cerebral hemispheres pared away dorsally notice the exposed corpus callosum consisting of tracts of fibres running directly from one hemisphere to the other.

In a brain with the cerebral hemispheres cut away still further (Fig. 36) notice the hippocampus. This is the posterior edge of the cortex curled under the rest of the hemisphere and lies as a transverse band just in front of the midbrain on either side. Below the hippocampus emerges the cerebral peduncle connecting the cerebral hemispheres and the midbrain. In front of the hippocampus lies the narrow ventricle of the cerebral hemisphere, with the corpus striatum forming its floor and outer anterior side. Notice the enormous development of the roof portion of the cerebral hemispheres (cerebral cortex) compared with the floor (corpus striatum). In lower Vertebrates the floor of the hemispheres is more bulky than the roof.

The functions of the different parts of the brain, and the paths of co-ordination between them should be noted and compared with those of other types. (See general section on nervous system, p. 153.)

12 *General account of the vertebrate nervous system*

THE nervous system serves to correlate the various activities of the organism. It is derived from ectoderm and the essential element is the nerve cell. The nerve-cell is divisible into two portions, the main body of the cell, the neuron, and usually one long process or fibre, the axon. The neuron contains the nucleus of the cell and the bulk of the protoplasm. In addition to the axon smaller processes, the dendrites, project from the surface of the neuron. In some nerves the axon is surrounded by the myelin or medullary sheath which is a development from the neuron. In the peripheral parts of the axon, i.e. outside the brain and spinal cord, it is covered by a delicate sheath, the neurilemma, enclosing the fibrillae and the myelin sheath.

In the Vertebrata the nervous system is concentrated to form the spinal cord which has an expansion at its anterior end, the brain. A nerve as dissected out consists of a number of axons bound together by a sheath of connective tissue. The neurons are situated along or close to the spinal cord. The axons in one nerve may come from neurons in different parts of the nervous system so that the one nerve may be transmitting impulses of several different kinds. The axons may be of extreme length, almost as long as the body of the animal.

Whereas the peripheral nerves consist of axons, the neurons aggregate together in the brain, the spinal cord, the ganglia on the dorsal roots of spinal nerves and on certain of the cranial nerves, and in the sympathetic ganglia. In this central nervous system, stimuli of a certain kind are directed to certain portions of it, no matter what their origin. Similarly impulses of similar nature are sent out from particular regions, no matter what their destination may be. This directness of the path along which the stimulus travels and the centralization of cross connections results in a gain in speed and efficiency.

The spinal cord, as being the least specialized portion of

the central nervous system, should now be investigated. The spinal cord consists of a tube with a very thin canal running down the centre enclosed by thick walls. Running down the dorsal and ventral surfaces will be seen median fissures. In a transverse section the spinal cord may be seen to be composed of an outer layer, the 'white matter', composed of medullated nerve-fibres, and the inner mass of 'grey matter', composed of neurons, with connective tissue, the neuroglia, between them. The grey matter lies surrounding the central canal and projects on either side as two longitudinal ridges, the dorsal and ventral cornua. In each segment the paired spinal nerves arise by a dorsal and a ventral root on either side leaving the cord opposite the cornua of grey matter. On the dorsal root close to its origin is a small ganglion. The dorsal root carries sensory (afferent) nerve-fibres and the ventral root, motor (efferent) fibres. In primitive forms, such as *Amphioxus* and *Petromyzon*, these roots run separately to the sense organs and muscles. In the higher forms the dorsal and ventral roots unite immediately after leaving the spinal cord and run together as a mixed nerve finally separating again peripherally to supply the sense organs and muscles. The motor fibres arise from neurons lying in the ventral cornua of grey matter. The axons from these neurons pass out with the spinal nerves to all parts of the body and end by branching processes, or more specialized end organs, upon the muscles. The sensory neurons lie in the ganglia on the dorsal root. These nerve cells are bipolar, i.e. the neuron bears two long axons. One of these axons passes outwards with the spinal nerve and ends in specialized sense organs or sense cells. The other axon passes inwards to the dorsal cornu of grey matter.

Besides the fibres forming the sensory and motor roots of the spinal nerves other motor fibres are connected to the spinal cord laterally. These belong to the autonomic nervous system, the greater part of which is composed of the sympathetic system. The autonomic nervous system is essentially an efferent system and transmits impulses to the stomach, intestine, and blood-vessels. The sympathetic fibres primitively entered and left the spinal cord via the dorsal root nerve but in the trunk of Mammals, the motor sympathetic fibres are all in the ventral root nerves. The special visceral motor system of the head (see later) continues to leave the central nervous system via the dorsal root nerves. These efferent sympathetic nerve-fibres are co-ordinated by the sympathetic ganglia quite independently from the brain and spinal cord, and thus independently of volition or consciousness. Afferent or sensory fibres in the sympathetic system are much less numerous than the efferent.

The sensory and motor nerve-cells entering and leaving the spinal cord are further connected to other nerve cells within the spinal cord. Thus each sensory axon entering by the dorsal root of the spinal cord forms a close association with another sensory nerve-cell. The end of the entering sensory axon branches considerably and spreads round the second neuron becoming closely associated with its dendrites. This union is known as the synapse and by it the nervous impulse is transmitted from one nerve cell to the next. The axon of this second nerve cell passes out directly into the white matter of the spinal cord or first forms a synapse with another nerve cell. In either case the impulse is carried forwards by an axon lying in the dorsal white matter of the spinal cord between the dorsal cornua of grey matter. All the axons in this region together form the ascending sensory tract of the spinal cord which runs forwards into the brain where certain sensory impulses may become conscious.

The white matter of the spinal cord on the ventral side between the ventral cornua transmits fibres originating from motor neurons lying in the brain. They pass down the spinal cord forming the descending motor tract. From this tract the fibres pass into the grey matter and there unite by synapses directly with the motor neurons of the ventral cornua. From these neurons the motor-fibres pass out to the ventral roots of the spinal nerves.

It is not necessary for all sensory impulses to pass forwards to the brain. Certain cross connections are formed between the sensory fibres as they enter the cord and the motor neurons in the ventral cornua. These connections are made by one or more intervening nerve cells. As a result of this, motor responses to certain stimuli may be effected without the stimulus first travelling to the brain. This short circuit mechanism is usually termed a reflex arc and plays a very large part in the control of the reactions of the body.

There are four main components of the central nervous system. These components are formed by nerves that are derived from special centres in the brain and spinal cord. The components are as follows.

(1) Somatic sensory nerves, conveying sensory impulses from the outside world, such as sensations of pain and touch.

(2) Visceral sensory nerves, conveying sensory impulses from the inner parts such as from the gut.

(3) Visceral motor nerves, conveying motor impulses to the smooth visceral muscles of the trunk region. The visceral motor nerve fibres do not run all the way from

the spinal cord to an organ but make synaptic connections with the neurons in the sympathetic ganglia.

(4) Somatic motor nerves, conveying motor impulses to the striped muscles of the body wall and limbs.

It should be remembered that a nerve fibre or axon transmits only one kind of impulse. This is shown clearly in the case of sensory nerves which transmit only the sensation of the end organ that they supply. In whatever way the nerve is stimulated it transmits the same kind of impulses.

The brain
(Figs. 8, 9, 15, 21, 36–8)

The neural tube in forms above *Amphioxus* is expanded at its anterior end. This expansion is produced by the dilatation of the central canal aided by the thickening of the walls in certain places. In the embryo the dilatation leads to the formation of three distinct swellings one behind the other, each with a well marked cavity or ventricle. These primitive divisions of the early brain are termed the forebrain, the midbrain, and the hind brain.

At the anterior end of the forebrain there are developed the paired olfactory lobes. These are concerned with the sense of smell and the part of the forebrain behind them, the cerebral region, is primitively occupied by correlation centres concerned with olfaction. In the lamprey, there is a single cavity with a small area of non-nervous roof in the cerebral region. In Teleost fishes, the non-nervous roof is greatly expanded and the nervous material is concentrated in two large masses on the ventral side. This type of forebrain is called 'everted'. In Elasmobranch fishes, the nervous material meets dorsally and there is a median pillar of nervous tissue dividing the cavity into two lateral ventricles. This is the 'inverted' type of forebrain. In Dipnoan fishes and in Tetrapods, however, there is the same arrangement as in the lamprey posteriorly but the forebrain cavity is enlarged forwards on either side of the lamina terminalis (the original anterior end) to form a pair of lateral ventricles, each completely surrounded by nervous tissue. This is the 'evaginated' type of forebrain. In all fishes, it is possible to recognize three different concentrations of neurons in the forebrain—these are the dorsal (archepallium), lateral (palaeopallium), and ventral (striatum) areas. In Tetrapods a fourth area is developed, the neopallium, between the dorsal and lateral areas. This becomes the principal correlation centre of the brain and in Mammals it has a very complex structure and is called the 'cerebral cortex'. This area is less well developed in Birds where the striatum is the largest correlation centre. In Teleost fishes, the main correlation centre is in the midbrain, in the optic lobes.

The posterior end of the forebrain, the diencephalon, has

a thin roof anteriorly, its cavity expanding in the adult brain to form the third ventricle which communicates with the lateral ventricles of the forebrain by the foramen of Monro. A portion of this thin roof is often highly vascular, forming the choroid plexus. The roof of the midbrain forms one or two paired enlargements called the optic lobes or corpora quadrigemina. Ventrally the diencephalon forms a downwardly projecting outgrowth, the tuber cinereum, and the cavity of this tuber is called the infundibulum. The thickening of the sides of the diencephalon forms the thalamus. The floor and sides of the midbrain are thickened to form the nerve tracts connecting the hindbrain with the forebrain. These tracts are usually visible as distinct elevations of the ventral surface forming the crura cerebri.

The anterior end of the roof of the hindbrain is enlarged to form a structure varying greatly in size in different animals and known as the cerebellum. In general this is large in active animals and small in those that are sluggish in behaviour. Behind the cerebellum the roof is thin. This thin roof forms a covering of the enlarged ventricle of the hindbrain, the fourth ventricle. At its upper lateral edges the hindbrain forms two swellings, the restiform bodies. The anterior portion of the ventral wall of the hindbrain is thickened and forms the pons. The pons is continued upwards on either side to form the middle peduncle of the cerebellum and connects the latter with the medulla. Posteriorly the hindbrain merges into the spinal cord.

The cranial nerves
(Figs. 8, 9, 15, 21)

In the head of the primitive Chordate the cranial nerves may be supposed to have been arranged segmentally and to have had their sensory and motor nerves completely separate as in *Amphioxus* and *Petromyzon*. In *Amphioxus* there is no marked distinction between the segments at the anterior end of the body and those of the middle and posterior regions, and the nerves are similar throughout. There is in fact no head. In the development of the head region the most anterior segments fuse together. The head is thus formed by the unsegmented anterior end of the body to which is fused a variable number of segments of the trunk.

In the head thus formed most of the cranial nerves can be traced as belonging to separate segments. They represent the dorsal and ventral nerve roots of each segment which have not united with one another immediately after their origin as have the spinal nerves. In the brain the nerve cells of each component are located in a special area, and the fibres of the nerves are all derived from these areas. The positions of origin of the cranial nerves within the brain, and the places of exit of these nerves, have been greatly altered owing to the extensive changes undergone by the anterior segments in forming the head. The olfactory and the optic nerves are not comparable with the segmental nerves, the former is produced by the nasal epithelium while the latter is an outgrowth of the brain. The following table shows the arrangements of the segments of the head in forms above the Cyclostomes, with the dorsal and ventral nerve-roots belonging to each.

HEAD SEGMENT	DORSAL ROOT	VENTRAL ROOT
premandibular	V profundus	III oculomotor
mandibular	V trigeminal	IV patheticus
hyoid	VII facial	VI abducens
4th segment	IX glossopharyngeal	no root
5th segment	X vagus (1st root)	no root
6th segment	X vagus (2nd root)	hypoglossal (1st root)
7th segment	X vagus (3rd root)	hypoglossal (2nd root)
8th segment	X vagus (4th root)	hypoglossal (3rd root)

The four components found in the spinal nerves are present in the cranial nerves, but both the somatic sensory system and the visceral motor is divided into two parts. The general somatic sensory system carries impulses to the brain of touch, pain, etc., and the acoustico-lateralis portion conveys to the brain impulses of hearing and equilibrial sense from the ear and lateral line sense organs. The visceral motor system consists of the parasympathetic system, in which the fibres are connected to the sympathetic ganglia in the same way as in the trunk system, and the special visceral motor system which supplies the striped muscles of the jaws and visceral arches.

The following shows the main division of the cranial nerves in Fish, according to their components. When the same nerve is placed in two or more systems it indicates that it contains fibres belonging to more than one system.

General somatic sensory system

V Superficial ophthalmic.
V Maxillary.
Small branches of IX and X (too small to be seen in dissection).
All the fibres of these nerves are derived from nerve-cells in the dorsal part of the medulla.

Acoustico-lateralis system

VII Superficial ophthalmic.
VII Inner and outer buccals.
VII External mandibular.

VIII Auditory.
   IX (Fibres too small to be seen in dissection.)
   X Vagus lateral line branch.
   The fibres of all these nerves are derived from nerve-cells in the acoustico-lateralis area of the medulla.

Visceral sensory system

  VII Palatine and prespiracular.
  VII Chorda tympani.
  VII Hyoidean.
   IX The visceral sensory fibres are inextricably mingled with the visceral motor fibres on the posterior border of the first gill cleft. The branch to the anterior border is purely visceral sensory.
   X The branches to the gill clefts are arranged similarly to IX, the anterior being purely visceral sensory and the posterior mixed.
   X Visceral branch to the stomach, etc.
   The fibres of all these nerves are derived from the visceral sensory tract in the medulla.

Special visceral motor system

   V Mandibular.
  VII Hyoidean.
   IX Branch to the posterior border of the first gill cleft.
   X Branches to the posterior borders of the remaining gill clefts.
   The fibres of all these nerves are derived from nerve-cells lying in the special visceral motor tract in the medulla.

Visceral motor system (autonomic)

  III Oculomotor.
   X Branches to the stomach, intestine, heart, and pharynx.

Somatic motor system

  III Oculomotor.
  IV Pathetic.
  VI Abducens.
 XII Hypoglossal.

   The fibres of III and IV are derived from nerve-cells in the midbrain, and the fibres of VI and the hypoglossal from nerve-cells in the medulla.
   The acoustico-lateralis system is only developed fully in Fish and less completely in Amphibia. In Reptiles, Birds, and Mammals it is reduced to the auditory portion alone. Thus of the lateralis components of the cranial nerves in the Fish only the VIII nerve persists in the Reptiles, Birds, and Mammals.
   The XI nerve (spinal accessory) which occurs in higher Vertebrates is formed of fibres separated from the X nerve

(vagus). In the frog it forms a portion of the vagus (see Fig. 21(a)).

The XII nerve (hypoglossal) in Fishes and Amphibia is formed from the first spinal nerve. In the Reptiles, Birds, and Mammals this nerve becomes intra-cranial and arises from the medulla.

Regions and tracts of the brain of the skate
(Figs. 8, 9)

Certain portions of the brain are definitely allocated for the reception or emission of particular impulses.

Olfactory impulses pass from the olfactory nerves to the olfactory tracts and so to the fore brain which is mainly concerned with smell. The infundibulum and inferior lobes of the midbrain are also concerned mainly with smell.

Optic impulses reach the brain by the optic nerves and pass along the optic tracts to the optic lobes which are mainly concerned with visual impulses.

The restiform bodies and paired thickenings of the upper part of the walls of the fourth ventricle together constitute the acoustico-lateral areas, or primary centres of the lateral line system. To these areas pass all the lateralis components of the cranial nerves.

Other nerve tracts can be seen in the floor of the fourth ventricle which pass to the spinal cord, the forebrain and the midbrain. They are inconspicuous in the skate but can be seen clearly in the spiny dogfish (*Acanthias*). The somatic motor tracts form two very prominent ridges running along the floor of the fourth ventricle. Just beside the somatic motor ridges lie the much less prominent visceral motor ridges indicating the visceral motor tract. Outside this again lies the visceral sensory tract. On the ridge forming this tract may be seen five swellings which indicate the origin of the nerves of each gill cleft. The somatic sensory tracts can be seen forming the dorsal part of the spinal cord. In the transverse section of the spinal cord at the base of the brain these tracts can be seen lying between the anterior cornua of the grey matter. Follow these tracts forwards and note that they form the upper posterior borders of the fourth ventricle running forwards at the side of the acoustico-lateral area.

Deeper tracts within the brain cannot be seen in surface view.

Regions and tracts of the brain of a sheep
(Figs. 36–8)

The higher Vertebrates are characterized by the development of a cerebral cortex from the roof of the forebrain. In the Fish the grey matter composed of neurons in the forebrain lies around the ventricles just as the grey matter in the spinal cord lies round the central canal. In the higher Vertebrates the neurons gradually wander outwards through the white matter and finally form a thick layer in the surface portions of the cerebral hemispheres as well as in the deeper parts of the fore-brain. This increasing outwandering of neurons is seen in the cerebral hemispheres of Amphibia and Reptiles and a fully formed cortex is seen in the Mammals.

The cerebral cortex is the seat of mental processes known as feeling and volition. Feeling or consciousness is the final phase of sensory stimuli, and volition is the starting-point of motor activity. The correlation of sensations with one another and with volitional impulses constitutes thought. The cerebral cortex is divisible into motor and sensory areas each associated with a different type of impulse.

The increasing development of the cerebral cortex both in size and area by the formation of superficial furrows is corre-lated with the higher mental capacity of the Mammals.

The main paths of impulses through the brain of the sheep are as follows.

Olfactory impulses pass along the olfactory nerves into the olfactory lobes which are the primary olfactory centres; they are then relayed along the olfactory tracts to the pyriform lobes, which form the secondary olfactory centres; from each pyriform lobe they pass to the hippocampus, the tertiary olfactory centre or conscious centre of smell. The hippo-campus has extensive connections with other parts of the brain for reflex purposes by way of the fornix which connects with the mamillary body, etc. In the dogfish practically the whole of the forebrain is concerned with smell, while in the sheep the olfactory functions occupy but a part, the remainder having developed new connections and functions.

Optic impulses travelling along the optic tracts reach the primary optic centre in the lateral geniculate body. From here they pass through the cerebral peduncle to the cerebral hemispheres and by the pulvinar of the thalamus, a visual centre of the diencephalon, to the conscious visual area in the cortex. The anterior corpora quadrigemina are reflex centres for optic impulses.

The primary auditory centre lies in the area acoustica, from here the impulses are carried in part by the trapezium to the posterior corpora quadrigemina (secondary centre) and medial geniculate body (tertiary centre), and then by the cerebral peduncle to the cortex of the cerebral hemispheres. The trapezium also unites the acoustic areas on the two sides of the medulla. In the dogfish the area acoustica receives the lateralis portions of VII, IX, and X as well as the VIII nerve. The cerebellum is a specialized derivative of the area acoustica.

Impulses of position from the semicircular canals, and from sense organs (proprioceptors) in the joints, etc., via the spinal

cord enter the cerebellum by the inferior peduncles or restiform bodies. Fibres are relayed from the cerebellar hemispheres to the midbrain and so to the cerebral hemispheres
by the cerebral peduncles.

Similarly other sensations such as pain, touch, etc., are
carried by definite paths to the thalamus from which they are
relayed to the cerebral cortex. The thalamus is the primitive
centre for the reception of afferent impulses, and is the principal one in the lower Vertebrates which have no cortex. In
Mammals the thalamus is an important cell-station for the
sorting and grouping of sensory impulses on their way to the
higher centres in the cortex.

Impulses of voluntary movements originate in the cerebral
cortex, pass through fibres in the cerebral peduncles that
appear on the ventral surface of the medulla as the pyramidal
tracts or pyramids, and then descend the whole length of the
cord and connect with motor neurons.

The midbrain is the region where most sensory impulses
are relayed to the cerebrum. It is the great co-ordinating
centre for the analysis of actions concerned in balance, for the
co-ordination of auditory sensations and visual movements,
etc.

The cerebellum is the main receiving organ for impulses of
position and is extensively connected with the midbrain and
cerebrum.

There are also several main connecting paths between
different parts of the brain. The trapezium has already been
mentioned. Impulses from the cerebrum pass to the cerebellum via the cerebral peduncles, crura cerebri, and pons to the
middle cerebellar peduncles. Fibres in the crura cerebri cross
over and supply the opposite sides of the cerebellum.

The corpus callosum forms a direct connection between the
two cerebral hemispheres, so that impulses from one to the
other no longer have to pass through the basal corpus striatum
as in lower vertebrates. The corpus callosum is only found in
the Eutherian Mammals. In the Marsupials no corpus callosum is present and the cortex of the two hemispheres is only
united via the anterior commissure and lower portions of the
forebrain. This results in a far less intimate association between the two hemispheres.

13 *The skull, visceral arches, and head muscles*

THE vertebrate skull in its simplest form consists of a cartilaginous neurocranium surrounding the anterior portion of the neural tube with the jaws and visceral arches suspended below it. This is the condition seen in the dogfish or skate and embryos of higher forms and is called the chondrocranium. The skull of bony fishes and all other forms has, in addition, membrane bones that lie outside the chondrocranium and are formed from dermal plates of bone. Thus the skull consists originally of three elements:

(1) cartilaginous brain-box (neurocranium),
(2) skeleton of the jaws and hyoid arch (splanchnocranium),
(3) dermal (membrane) bones.

The neurocranium is usually incompletely roofed and includes the paired sense-capsules of the nose and ear at its front and hind end respectively. A complete ring of cartilage surrounds the foramen magnum by which the spinal cord leaves the brain box. There is usually a large orbit that protects the eye; the sclerotic layer of the eye may contain cartilage or bony plates.

The chondrocranium (cartilaginous neurocranium plus splanchnocranium) is more or less ossified in most adult vertebrates; the bones which replace the cartilage are called *cartilage bones*. In addition, *dermal (membrane)* bones form under the skin and in the roof of the mouth. Primitively these lie external to the cartilage bones, separated from them laterally by the jaw and branchial muscles and dorsally and ventrally by connective tissue. In more advanced vertebrates, the original heavy roofing of dermal bones is reduced and these bones sink under the skin so that they lie close to the cartilage bones. Some composite bones are formed and various bones, of either origin, may be absent. Changes in the arrangement of the bones are associated with better development of the brain and sense organs and with increased efficiency of feeding.

Skull of the dogfish or skate

In the dogfish and the skate the skeleton of the jaws and gill arches are attached by ligaments only to the neurocranium.

Dorsal surface of the neurocranium

At the anterior end of the skull are the large cartilaginous olfactory capsules surrounding the olfactory lobes. They are separated from each other by a median internasal septum. In the dogfish there are three rostral cartilaginous bars projecting forwards which support the snout. The two outer of these project from the olfactory capsules while the middle one is a prolongation of the floor of the cranium itself. In the skate the rostrum is a very large and prominent structure. The roof of the skull is incomplete in front forming the anterior fontanelle. In the dogfish the roof of the skull is complete behind the anterior fontanelle. In the skate in front of the auditory capsules is a large oval space forming the posterior fontanelle. The posterior fontanelle is continuous with the anterior fontanelle so that the major portion of the cartilaginous roof of the skull is missing in the skate. The sides of the cranium are drawn out into the supra-orbital crests forming the dorsal boundaries of the orbits. The auditory capsules lie at the lateral posterior sides of the cranium and are completely fused with it. The front of the auditory capsule forms the posterior part of the orbital cavity. In the dogfish, between the two auditory capsules there is a median depression in the roof of the cranium. At the bottom of this depression is a pair of foramina leading downwards and opening to the auditory capsule. These foramina carry the paired endolymphatic ducts. In the skate the two foramina of the endolymphatic ducts open separately on the dorsal side of the cranium between the auditory capsules.

Ventral surface of the neurocranium

Anteriorly may be seen the paired olfactory capsules, which are completely open ventrally and are separated by the internasal septum. The floor of the cranium is complete and flat. In the dogfish it is drawn out laterally into flanges forming part of the floor of the orbital cavity. These flanges are absent in the skate.

Posteriorly between the auditory capsules the cranium narrows to form the foramen magnum through which the brain unites with the spinal cord. On either side of the foramen magnum lie the occipital condyles. The inner walls of the orbit are pierced by foramina for the passage of nerves and blood-vessels.

The visceral skeleton consists of a series of crescentic carti-lages lying in between the gill slits in the pharyngeal region. In considering the visceral skeleton it must be borne in mind that the mouth and the spiracle and the gill clefts are homo-logous.

The posterior branchial arches are the least modified, while of the anterior ones the skeleton of the jaws is the most modified; the premandibular arch is completely incorporated into the neurocranium.

Select for examination one of the five posterior branchial arches. In the dogfish each branchial arch is formed of four elements on each side. The most dorsal of these elements lying nearest to the vertebral column is the pharyngo-branchial. Next to the pharyngo-branchial comes the epi-branchial. Next to this comes the cerato-branchial and below it the hypo-branchial; the last gill arches in the dogfish have no hypo-branchials.

The basi-branchial is median and unpaired and connected with the third and fourth hypo-branchials in front, and behind with the fifth cerato-branchials.

Both the epi-branchials and the cerato-branchials, except those of the fifth arch, bear gill rays that support the gill filaments.

In the skate the pharynx is dorsoventrally flattened and the basi-branchial forms a large median plate, the copula. There is only one pair of hypo-branchials which articulate with the copula and with the second and third cerato-branchials. All the cerato-branchials overlap ventrally and the fifth articulates with the copula. It is difficult to see the divisions in the dorsal parts of the visceral arches.

The branchial arch immediately behind the jaws is termed the hyoid arch and is larger than the branchial arches behind it and differs from them slightly in structure. It is joined to the skull by ligaments and to it are attached also by ligaments the upper and lower jaws. The most dorsal portion, the hyo-mandibular, is attached to the skull by a ligament. There is a special facet on the auditory capsule provided for the articu-lation of this cartilage with the skull. Note the small hole in the hyomandibular, for the passage of an artery. The next portion of the arch is the cerato-hyal; in the dogfish the cerato-hyals abut on the median basi-hyal in the middle line; the epi-hyals and the hypo-hyals are absent. In the skate the cerato-hyal is attached to the dorsal end of the hyomandibular by a small epi-hyal, the hypo-hyals are absent, and the basi-hyal is a narrow cartilage lying behind and parallel with the lower jaw.

The jaws represent a pair of branchial arches very considerably modified. In each two pairs of cartilages are present, and at the upper and lower ends these arches unite together in the middle line. The upper jaw consists of paired cartilaginous bars called the palato-pterygo-quadrate bars. The reason for this long unwieldy name being applied to a smooth cartilaginous bar is because in bony skulls it appears to ossify from three centres to form the palatine, the pterygoid, and the quadrate bones. The lower jaw closely resembles the upper jaw. The two cartilages of which it is formed are known as Meckel's cartilages.

The jaws are attached to the skull and to the hyoid arch by ligamentous connections only. The upper jaw is attached to the skull by the ethmo-palatine ligament connecting the anterior end of the upper jaw with the posterior border of the nasal capsule. At the angle of the jaws there is a ligament connecting both the upper and lower jaws to the junction of the hyomandibular and cerato-hyal cartilages.

This indirect suspension of the posterior end of the jaws is called 'hyostylic suspension'.

Teeth

The teeth in the skate form a pavement of broad flat denticles covering the lips and are used for crushing and rasping. The teeth represent modified placoid scales and are embedded in the fibrous integument covering the jaws and are not attached to the skeleton. They are developed from a groove in the mucous membrane of the mouth just within the jaws. Here new teeth are successively formed, and, as they grow, the skin shifts outwards over the jaws bringing rows of teeth to the edge, where they take the place of their predecessors which shift outwards and drop off.

Muscles of the dogfish head
(see pp. 11–14 for the skate)

There are three sets of visceral muscles related to each branchial arch.

(1) The *levators* brace the epi-branchials against the dorsal myotomes.

(2) The *constrictors* run dorsoventrally and close the gill slits.

(3) The *transverse ventral* muscles lie below the hypo-branchial muscles (described later) and compress the branchial basket causing expiration of water through the gill slits. This contraction is the active movement of respiration and when the *transverse ventral* muscles relax, the branchial region expands as a result of elasticity of the arches and so water is inspired, mainly through the mouth.

The hypo-branchial muscles (*coracobranchiales*) arise as a mass, the *arcuales communis*, attached to the pectoral girdle ventral and lateral to the pericardium, and are inserted on the hypo-branchial cartilages. They may assist inspiration by lowering the floor of the pharynx but are probably more concerned with swallowing.

The muscles of the hyoid and jaw arches are somewhat modified. The *levators* brace the hyomandibular and palato-pterygoquadrate bar respectively against the neurocranium. The *constrictor hyoidei* controls the spiracular valve. The constrictor of the mandibular segment is the *adductor mandibulae*, running between the upper and lower jaws—this closes the mouth. The *transverse ventral* muscles resemble those of the gill segments. The hypo-branchial muscles insert on the hyoid (pair of *coracohyoideus* muscles) and on Meckel's cartilage (*coracomandibularis*). The latter opens the mouth.

There are no lateral plate or ventral myotomic muscles related to the premandibular segment.

Skull of a bony fish *(Gadus)* (Fig. 39)

It is better to use the skull of a cod for examination as it is much larger than that of the whiting.

The skull of the cod consists principally of bone. There is little of the chondrocranium left as cartilage. The bone arises in two ways: (1) by the ossification from separate centres of the chondrocranium; (2) by the formation of membrane bones in the tissue surrounding the chondrocranium. Some bones are formed by both these processes and are thus mixed in origin.

In order to see the relationship of this skull with that of the dogfish or skate it will be best first to look for those bones that have been derived by the ossification of the cartilaginous chondrocranium.

Bones derived from the cartilaginous neurocranium

Find first of all the cavity that contains the brain and thus identify the position of the neurocranium. The neurocranium round the foramen magnum is ossified to form four bones. The dorsal bone is the supraoccipital. This is a large bone and the dorsal surface of it is raised into a crest that projects upwards and backwards.

The foramen magnum is bordered on either side by the paired exoccipital bones. The exoccipitals are irregular shaped bones and each carries a foramen for the exit of the X nerve. The lower border of the foramen magnum is formed by the basioccipital bone. The posterior face of this bone forms the single concave occipital condyle. Anteriorly the basioccipital projects forwards in the middle line between the ends of the parasphenoid.

The auditory capsule is formed of five bones. The largest of these bones is the opisthotic, which is formed from the posterior ventral part of the auditory capsule and lies just

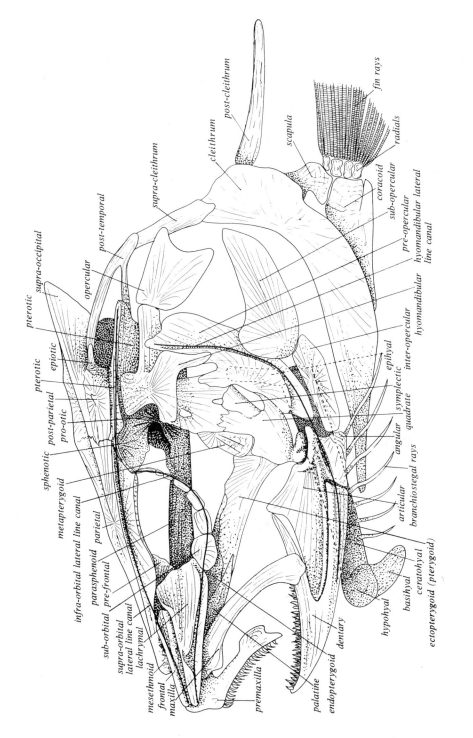

Fig. 39. Lateral view of the skull and pectoral girdle of the cod. The gill arches are not shown. The parts of the hyoid arch lying behind other bones are shown by dotted lines. The course of the lateral line canals along the head are shown by three black lines.

below the exoccipital. It has a foramen for the exit of the IX nerve. In front of the opisthotic can be seen the smaller pro-otic bone. The pro-otic lies at the lower anterior border of the auditory capsule and bears a notch across its anterior face which carries the V and VII nerves. The pterotic bone lies along the dorsal side of the auditory capsule, projecting as a wing from the upper posterior side of the skull. In front of the pterotic lies the sphenotic bone, which is a small bone just visible on the dorsal side in front of the pterotic, and running down ventrally to meet the pro-otic. The epiotic bone lies on the median side of the pterotic and abuts against the anterior lateral edge of the exoccipital. The sphenotic and the pterotic bones are not entirely derived from cartilage. They are partly formed from the surrounding membrane and are therefore bones of mixed origin.

The olfactory capsules ossify only in the middle line to form the median mesethmoid bone. The rest of the capsules remains cartilaginous.

The remainder of the neuro-cranium is reduced in size and, being cartilaginous, is not seen in the dried skull.

Skeleton of the jaws and visceral arches

The upper jaw, the palato-pterygo-quadrate bar, ossifies from three centres, the anterior of which forms a part of the palatine bone, the median, a part of the pterygoid, the posterior, the quadrate bone. The easiest way to identify these bones is to start at the posterior end. The quadrate is identified by its forming the articular facet for the lower jaw. The pterygoid projects forwards from the quadrate. There are in the cod three bones occupying a 'pterygoid' position between the quadrate and the palatine, viz. the metapterygoid, the ecto-pterygoid, and the endopterygoid. Of these only the metapterygoid is formed by ossification of a part of the palatopterygo-quadrate bar, the ectopterygoid and endo-pterygoid are membrane bones. The metapterygoid abuts on the dorsal anterior border of the quadrate. Joining the quad-rate to the palatine are the ectopterygoid and endopterygoid bones. As the names imply, the ectopterygoid lies outside the metapterygoid and the endopterygoid on the inside towards the middle line. The palatine lies along the outer anterior border of the endopterygoid and projects forwards as far as the nasal capsules, lying on the outer side of the mesethmoid bone. The palatine bone in the cod, and most higher forms, is not a true cartilage bone, but is partially formed from mem-brane bone as well; it is therefore a bone of mixed origin, like the sphenotic and pterotic bones.

Meckel's cartilage ossifies posteriorly to form the large

articular bone that articulates with the quadrate. The remaining bones of the lower jaw are membrane bones.

The jaw suspension of the cod is methyostylic and differs
from that of the dogfish in that there is an anterior articulation
between the palatine bone of the upper jaw and the ethmoid
region of the neurocranium.

The hyomandibular is completely ossified to form one large
and one small bone, the hyomandibular and the symplectic.
As in the dogfish or skate, the hyomandibular articulates with
the facet on the lateral upper border of the auditory capsule.
The hyomandibular is a large bone of irregular shape, narrow
at the point of articulation with the skull and extending forwards, backwards, and downwards. If the sutures between the
quadrate, the metapterygoid, and the hyomandibular are
traced it will be seen that there is a fourth bone, the symplectic,
lying between the quadrate and the hyomandibular and behind the metapterygoid. The symplectic extends along the
inner side of the quadrate and bears a facet for articulation
with the lower jaw. Posteriorly behind the quadrate, the
symplectic, and the hyomandibular lie the opercular bones.
These are membrane bones. The hyomandibular has a marked
opercular process on which the opercular bone is hinged.

The remainder of the hyoid arch is best seen from the inside.
It ossifies to form the following bones. The epiphyal is a small
bone projecting vertically downwards from the ventral border
of the hyomandibular. The cerato-hyal, the largest bone in
the arch, projects forwards at right angles from the ventral
border of the epihyal. At its anterior end is a small bone, the
hypo-hyal, which is joined to its fellow on the opposite side
by a ligamentous connection. Projecting downwards from
between the two hypo-hyals is a single median bone, the
basi-hyal.

Each element of the remainder of the visceral arches ossifies
separately to form the pharyngo-, epi, cerato-, and hypo-
branchial bones. The pharyngo-branchials of the anterior
arches are fused together. The hypo-branchials of the three
anterior branchial arches articulate with the median ventral
basi-branchial. The fused pharyngo-branchials and the ventral elements of the last arch bear teeth. The cerato-hyal bears
a number of backwardly projecting bony rods from its lower
posterior border, the branchiostegal rays, which support the
branchiostegal membrane. These rays are membrane bones.

Except where specific mention is made of the fact, all bones
so far described are derived from the ossification of the cartilaginous neurocranium and visceral arches. The remainder of
the bones of the skull are membrane bones that are formed by
ossifications in the connective tissue surrounding the neuro-

cranium. The appearance of the membrane and the cartilage bones in the fully developed skull is exactly similar and they cannot be distinguished from each other by inspection. In order to determine which is which a knowledge of the development of the skull is necessary.

Membrane bones of the skull

There is still doubt about the homologies between some of the membrane bones of Actinopterygian fishes and those of the Crossopterygians and Tetrapods. In the account that follows, the names used in the third edition of this book have been retained but it should be noted that in many other text-books the usage is different and the following are equivalent names for bones.

Name used here	*Name often used in other texts*
Post-parietal	Parietal
Parietal	Frontal
Frontal	Nasal

Working from behind forwards the following are the membrane bones of the roof of the skull (this roof of membrane bones is complete and the fontanelles of the neurocranium are covered over). On either side of the median supra-occipital bone lie the post-parietal bones. These extend outwards to the pterotics and backwards to the epiotics. In front of the post-parietals is the largest bone of the roof of the skull, the fused parietals. At the anterior border of the parietals there projects forwards in the middle line, the mesethmoid bone, a cartilage bone that has already been described. On either side of this and loosely joined to the parietals are the paired frontals. Outside the frontals lie two large bones, the lachrymals.

The upper jaw, which in the dogfish or skate was formed by the cartilaginous palato-pterygo-quadrate bar, is replaced functionally by two paired membrane bones. The arrangement of these is different in various groups of Teleost fishes. In the cod, the more anterior one, the premaxilla, bears teeth on its lower (anterior) border and is united dorsally to the premaxilla on the other side forming a short premaxillary spine posteriorly. Ligaments run from the premaxillae on each side to the palatines. The maxilla is a long slender curved bone posterior to the premaxilla. Its upper end is held by a ligament to the ethmoid region and by strong connective tissue to the anterior end of the palatine bone, and a thick fold of skin runs between its lower end and the articular bone of the lower jaw. The maxilla is toothless. When the cod opens its mouth, the posterior, lower ends of the maxillae and premaxillae swing forwards so that the gape is circular. In primitive Teleosts such as the trout, the premaxilla is short and lies in series with the maxilla; both premaxilla and maxilla bear teeth. Many teleosts protract the premaxillae when the mouth opens and there are several different anatomical arrangements allowing this movement. Characteristically, the premaxillae lie in front of and parallel to the maxillae, as in the cod, and usually there is a long premaxillary spine.

The membrane bones along the side of the skull do not make such a complete investment as do the dorsal roofing bones. In a side view the following membrane bones will be seen besides the premaxilla, the maxilla, and the lachrymal bones that have already been described: in front of the frontal bone and somewhat overlain by its anterior border lies the prefrontal bone. This prefrontal bone together with the lachrymal, form the anterior border of the orbit. The upper border of the orbit is formed by the edge of the frontal bone. The lower and posterior borders of the orbit are formed by a chain of small sub-orbitals which are usually removed in the preparation of the skeleton.

The endopterygoid and ectopterygoid have already been noticed (p. 166). In the posterior region of the skull, behind the hyoid arch, lie the opercular bones. The largest and most anterior of these articulating with the quadrate and hyomandibular, is the preopercular. Below the preopercular lies the interopercular and behind it the subopercular. The opercular lies above the subopercular.

Here it will be necessary to notice the post-temporal bone behind the auditory region of the skull. This is a V-shaped bone and is usually left attached to the pectoral girdle in the preparation of the skeleton. The point of the longer arm of the V is attached to the backward process of the epiotic bone, while the point of the shorter arm articulates with the backward process of the pterotic bone. The apex of the V articulates with the supra-cleithrum.

The lower jaw is formed by the ossification of Meckel's cartilage to produce the articular bone, already described, together with the following membrane bones. The large paired dentary completely invests the cartilaginous anterior end of Meckel's cartilage. It bears teeth and is joined at its anterior end with its fellow in the middle line. Behind the articular bone lies the small angular bone.

The floor of the neurocranium is almost completely invested by membrane bones. The only cartilage bone which takes any part in the formation of the floor is the basioccipital, which has already been noticed. The largest bone on the floor of the skull is the parasphenoid. Posteriorly this bone diverges to form the V enclosing the forward projection of the basioccipital. It extends upwards to meet the otic bones and its anterior end lies against the prefrontal bones. The vomer is a single median bone lying at the anterior end of the floor of the cranium. It projects forwards between two processes of the parasphenoid which diverge anteriorly in the same way as at its posterior end. On its anterior border the vomer bears teeth.

TABLE SHOWING THE DERIVATION OF THE BONES OF THE SKULL OF THE COD	CARTILAGE BONES	MEMBRANE BONES	MIXED BONES
	Neurocranium		
	Exoccipitals	Frontal	Prefrontal
	Supraoccipital	Parietal	Pterotic
	Basioccipital	Post-parietal	Sphenotic
	Pro-otic	Parasphenoid	
	Opisthotic	Vomer	
	Epiotic	Lachrymal	
	Mesethmoid	Postorbitals	
	Basisphenoid	Suborbitals	
	Splanchnocranium		
	Metapterygoid	Premaxilla	Palatine
	Quadrate	Maxilla	
	Symplectic	Ectopterygoid	
	Articular	Endopterygoid	
	Hyomandibular	Dentary	
	Bones of hyoid arch	Angular	
	Bones of branchial arches	Opercular bones	
		Branchiostegal rays	

Teeth

Numerous sharp conical teeth are present on the premaxilla, dentary, and the anterior end of the vomer. They lie close together in a mass. Some teeth are fused to the bones that bear them by a bony substance which is reabsorbed when the tooth is shed. Some teeth are movably attached to the jaws by an elastic ligament on the inner side, allowing them to be folded back when food is taken into the mouth. New teeth are developed throughout life and arise behind the functional teeth. These new teeth take on the function and position of the old teeth as these are worn down and cast off.

Muscles of the cod's head
(Fig. 40)
The mechanical structure of the cod's skull

The bony skull of a Teleost fish can be divided into the following movable parts which can all be identified in a fresh cod's head.

(1) The *neurocranium* forms a brain box with a dorsal roof and articulates posteriorly with the vertebral column.

(2) The *palatal complex* consists of palatine, pterygoids, quadrate, symplectic, hyomandibular, and preopercular on each side, and articulates with the neurocranium

at two places. The anterior articulation is between the palatine and the ethmoid region, and the posterior one is between the hyomandibular and the sphenotic. The shape of the joints is such that the palatal complex can only swing outwards and inwards.

(3) The *mandible* includes the dentary, angular, and articular, the latter articulating with the quadrate; the anterior end of the mandible can be raised and lowered about this fulcrum.

(4) The *maxilla* is hinged at its upper anterior end to the palatine and is held by a ligament to the ethmoid, and the lower end is free to swing forwards and backwards. Another ligament connects the upper end of the maxilla to the mandible.

(5) The *premaxilla* is joined mid-dorsally to the other premaxilla and is hinged to the ethmoid region immediately anterior to the maxilla. It is held by a ligament to the palatine and the lower end can swing forwards and backwards in front of the maxilla.

(6) The *operculum* consists of opercular, subopercular, and interopercular, and is hinged to the opercular process of the hyomandibular and is held by strong connective tissue to the preopercular. The operculum can move towards and away from the body. The interopercular is attached by a ligament to the articular behind the quadrate articulation.

(7) The rest of the *hyoid arch* (i.e. epi-hyal, cerato-hyal, hypo-hyal, and basi-hyal) is articulated with the symplectic through the epi-hyal, and can be swung backwards and forwards about this joint.

Associated with the skull are the branchial arches and the pectoral girdle. The *cleithral arch* of the latter forms a firm posterior border to the gill region and is the anterior attachment of the ventral body muscles. The whole girdle can swing backwards and forwards about the articulations between the post-temporal and the pterotic and epiotic bones.

The dissection of the muscles

If a fresh cod's head is available, hold it firmly by the dorsal surface, grasp the pelvic fins and pull them backwards. The cleithral arch, hyoid arch, and mandible all swing backwards, the mouth opens wide, and the branchial chamber is dilated. Skin the ventral surface to expose the *geniohyoideus* and *sternohyoideus* muscles which produce in the living fish the movements observed above. They arise respectively on the basihyal and cleithrum and insert on the mandible and basihyal respectively.

Skin one side of the head and remove the postorbital, suborbital, and lachrymal bones and the eyeball, noting the four rectus and two obliquus muscles. Remove the superficial connective tissue to expose the *adductor mandibulae* (1), a small slip of muscle inserted on to the maxillo-mandibular ligament. Clear the length of this ligament and observe its dorsal and ventral insertions. Reflect adductor (1) forwards from its origin on fascia and dissect out *adductor mandibulae* (2), a large muscle arising from the outer surface of the ptery-goid, symplectic, quadrate, and preopercular, and inserting by a strong tendon on to the mandible and on to the tendon of the *intramandibularis*, a muscle that lies in a groove on the inner side of the lower jaw. Expose this muscle by dissection from the ventral side. Reflect adductor (2) ventrally and dissect out *adductor mandibulae* (3), which arises on the pterygoid and inserts on the maxilla below the articulation of this bone with the palatine. Reflect this muscle forwards to expose *adductor mandibulae* (4), which arises from the anterior face of the hyomandibular and inserts on to the tendon of adductor (2). Pull on each of these adductor muscles and observe their actions: adductors (1) and (3) retract the maxilla, while adductors (2) and (4), together with the intramandibu-laris, raise the mandible and thus close the mouth. Note that the premaxilla moves with the maxilla, and that the lower end of the maxilla is drawn forwards as the mouth opens because of the strong fold of skin that connects the dentary with the maxilla. Remove the frontal (nasal) bone and expose the 'crossed ligaments' that connect the maxilla with the ethmoid and the premaxilla with the palatine.

Dissect out the *levator hyoidei* muscle which arises from the sphenotic ridge and inserts on the outer surface of the hyo-mandibular. This muscle swings the palatal complex out-wards. Now expose the *adductor arcus palatini* which lies medial to the adductores mandibulae and arises from the parasphenoid and inserts on the pterygoid and internal surface of the quadrate. This swings the palatal complex inwards and is the antagonist of the levator hyoidei.

Dissect out four muscles connecting the operculum with the otic ridge. From anterior to posterior, these are the *dilator*, the *adductor*, and the two *levatores operculi*. The dilator is inserted on to the outer surface of the operculum, while the other three are inserted on its inner surface. Their names describe their actions but *levatores operculi* when they pull the opercular bones upwards also open the mouth slightly because of the ligament between the interopercular and articular. Remove the skin from the branchiostegal membrane and observe the constrictor muscles between the branchio-stegal rays.

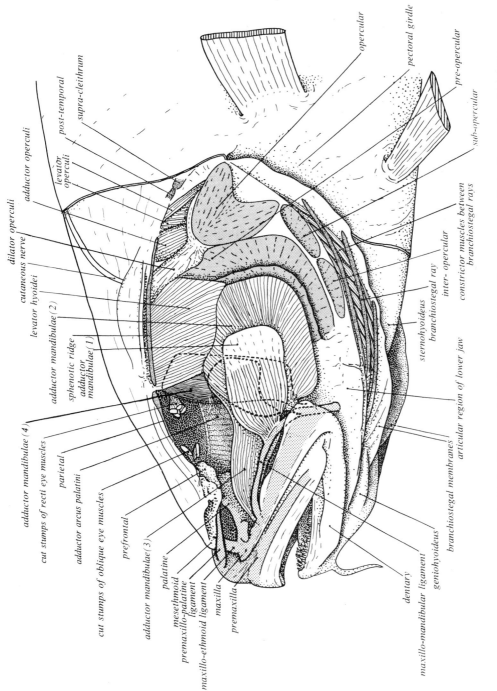

dilator operculi

cutaneous nerve

adductor operculi

levator operculi

post-temporal

supra-cleithrum

opercular

pectoral girdle

pre-opercular

sub-opercular

levator hyoidei

adductor mandibulae (2)

sphenotic ridge

adductor mandibulae (1)

constrictor muscles between branchiostegal rays

inter-opercular

branchiostegal ray

sternohyoideus

adductor mandibulae (4)

cut stumps of recti eye muscles

parietal

adductor arcus palatini

cut stumps of oblique eye muscles

prefrontal

adductor mandibulae (3)

palatine

mesethmoid

premaxillo-palatine ligament

maxillo-ethmoid ligament

maxilla

premaxilla

dentary

maxillo-mandibular ligament

geniohyoideus

branchiostegal membranes

articular region of lower jaw

Fig. 40. Head of the cod with the skin removed and the connective tissue dissected away to display the muscles, which have been left in their natural positions. Where muscles overlap their margins are shown by dotted lines, and where bones are exposed, either partly or completely, they are shown by mechanical tint.

The sequence of events during respiratory movements is probably as follows.

(1) Inspiration

The operculum is held against the body by adductor and levator operculi and the branchiostegal constrictors. The mouth is opened slightly by levator operculi and widely by sternohyoideus. The maxilla is automatically rotated forwards carrying the premaxilla with it. The buccal cavity is enlarged by levator hyoidei.

(2) Expiration

The mouth is closed by adductores mandibulae (2) and (4); and the maxilla is retracted, pulling the premaxilla with it, by adductores mandibulae (1) and (3). The operculum is opened away from the body by dilator operculi. The buccal cavity is reduced in volume by adductor arcus palatini.

The muscles used in inspiration are innervated by hypoglossal nerves (geniohyoideus and sternohyoideus) or hyoidean branches of VII; those used in expiration are innervated by mandibular branches of V.

In feeding, the mouth is opened more widely than in respiration. Movements of the pectoral girdle, together with a rotation of the neurocranium upwards by bending of the anterior part of the vertebral column, may assist in this opening.

Tetrapod skulls and cranial muscles
(Figs. 41–3)

In the skulls of primitive bony fishes (e.g. *Polypterus* and *Osteolepis*), early Amphibia (Stegocephalia) and primitive Reptiles (Cotylosauria), there is a complete covering of bones over the temporal region and behind the orbit. This type of skull in Reptiles is called 'anapsid' and is found in modern Chelonia (see p. 180). The muscles that close the mouth, the *adductores mandibulae*, lie between the neurocranium and the roofing bones over the cheek and temporal region so that the volume, and hence the power, of these muscles is limited by the space available there. In the evolution of various groups of Reptiles from the Cotylosauria, this covering of dermal bones is reduced and the muscles have more room when they contract; they become diverse and often very powerful. In most Reptiles, 'vacuities' appear as the result of two or more bones failing to meet along their whole edges; these vacuities arise in different positions in different groups.

The *adductores mandibulae* of Reptiles, innervated by cranial nerve V, are divided into three parts (see Fig. 42):

(1) the *adductor externus* inserts on the outer side of the mandible and arises from the medial surface of the lower part of the temporal roof,

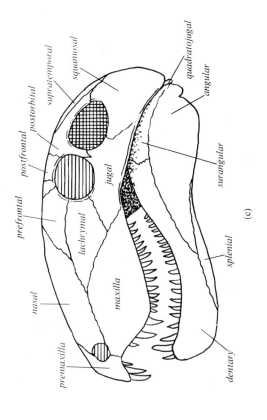

premaxilla

maxilla

lachrymal

prefrontal

jugal

nasal

frontal

parietal

postorbital

postfrontal

intertemporal

squamosal

supratemporal

tabular

quadratojugal

pterygoid

quadrate

posttemporal

postparietal

(a)

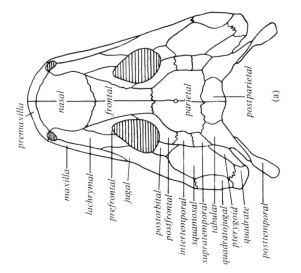

supratemporal

quadratojugal

quadrate

squamosal

postorbital

surangular

postfrontal

jugal

prefrontal

angular

nasal

lachrymal

premaxilla

maxilla

dentary

splenial

(b)

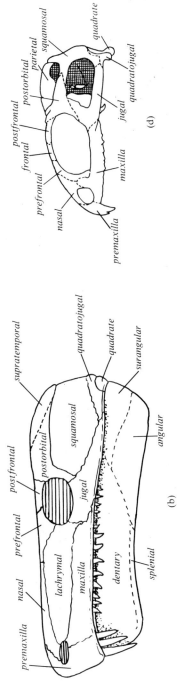

quadratojugal

angular

supratemporal

squamosal

postfrontal

postorbital

prefrontal

nasal

premaxilla

maxilla

lachrymal

jugal

surangular

splenial

dentary

(c)

quadrate

parietal

squamosal

postorbital

postfrontal

frontal

prefrontal

nasal

premaxilla

maxilla

jugal

quadratojugal

(d)

F<small>IG</small>. 41. Skulls of Amphibia and Reptilia. (a) Skull of *Palaeogyrinus*, one of the embolomerous Amphibia. (b) Skull of *Limnoscelis*, a cotylosaurian reptile showing an anapsid skull in side view. (c) Skull of *Dimetrodon*, a synapsid reptile, in side view. (d) Skull of *Sphenodon*, a diapsid reptile, in side view.

(2) the *adductor internus pars pseudotemporalis* inserts on the inner side of the mandible and arises from the medial surface of the upper part of the temporal roof and from the side of the neurocranium,

(3) the *adductor internus pars pterygomandibularis* (also called *pars pterygoideus* sometimes) arises from the lateral surface of the pterygoid and inserts on the medial and lower surface of the posterior end of the mandible (the angular bone).

These muscles all shut the mouth. The *pterygomandibularis* may also move the upper jaw in relation to the brain case (see later).

In the SYNAPSIDA there is a single vacuity between the jugal, postorbital, and squamosal bones and the *adductor externus* muscle is exposed on the cheek. This muscle arises from all round the vacuity and from the lower border of the jugal and quadratojugal. The mammals are descended from this group of Reptiles.

In the DIAPSIDA a second vacuity is present between the parietal, postorbital, and squamosal bones. This vacuity lies dorsal to that described above in the Synapsida and the two are known as the upper and lower temporal vacuities respectively; they are separated by the postorbital-squamosal arcade. The *adductor externus* muscle is exposed by the lower temporal vacuity and the *pseudotemporalis* muscle is exposed by the upper temporal vacuity. Modern diapsids include the Rhynchocephalia and Crocodilia, which exhibit the vacuities in unmodified form, and the Squamata, in which the vacuities are modified by the loss of the lower arcade (the connection between jugal and quadratojugal) in lizards and the additional loss of the upper arcade (the connection between postorbital and squamosal) in snakes. The Aves (Birds) are descended from diapsids.

Single temporal vacuities were present in several extinct groups of Reptiles. From the position of these in Euryapsida (Synaptosauria) and Ichthyosauria, it seems probable that only the *pseudotemporalis* muscle would be exposed in these animals. The vacuity of the Mesosauria resembled that of Synapsida.

Cranial kinetism
(Fig. 42)

A skull is described as 'kinetic' if all or part of the upper jaw is movably articulated with the brain case—cranial kinetism is thus a very ancient and basic feature of vertebrates and is found in almost all modern fishes but not in the Dipnoi. The earliest tetrapods probably had kinetic skulls and several types of kinetism can be observed among modern Diapsida. The functions of cranial kinetism possibly vary from species to species but the phenomenon is characteristic of carnivorous Tetrapods with deep skulls.

METAKINESIS is found in many lizards, e.g. *Lacerta*, and is probably a primitive condition. The skull consists of two portions, the occipital and maxillary 'segments'. The posterior part of the neurocranium, including the auditory capsules, forms the occipital segment and this supports the brain and articulates with the vertebral column. The rest of the skull bones, excluding the lower jaw and hyoid and branchial elements, form the maxillary segment. This can move about the occipital segment with the fulcrum posteriorly at the contact between the exoccipital, supratemporal, and squamosal and with sliding contacts dorsally between the supraoccipital and parietals and ventrally between the pterygoids and the basipterygoid process of the basisphenoid (or parasphenoid). Muscles running between the two segments and moving the upper jaw relative to the brain case are derived from the *constrictor dorsalis*, innervated by cranial nerve V. In lizards there are usually present a *levator pterygoidei*, arising from the pro-otic and the ventral flange of the parietal, running parallel to the epipterygoid bone and inserting on the dorsal surface of the pterygoid bone medial to the epipterygoid, and a *protractor pterygoidei*, arising from the pro-otic and basisphenoid and inserting along the dorsal surface of the whole length of the quadrate process of the pterygoid bone. The *levator bulbi*, a sheet of muscle under the eyeball, is derived from the same embryonic mass and has an origin from the pro-otic and an insertion on to the palatine process of the pterygoid. The *adductor internus pars pterygomandibularis* also may move the upper jaw relative to the brain case provided that the skull is also streptostylic (see next paragraph).

The lower temporal arcade is absent in all lizards and the quadrate is free to move about its articulation with the squamosal. At its lower end it is connected by ligaments to the quadrate process of the pterygoid and to the jugal. This condition is called STREPTOSTYLY. The opposite condition, where the quadrate is firmly sutured to the squamosal and through the quadratojugal to the jugal, e.g. in *Sphenodon*, is called MONIMOSTYLY.

In some lizards, e.g. *Varanus*, there is a dorsal joint between the frontals and parietals—this is the MESOKINETIC hinge. *Varanus* also has the metakinetic joint between the parietals and supraoccipital and therefore exhibits AMPHIKINESIS. Its skull consists of three 'segments' since the maxillary segment of *Lacerta* is represented here by a parietal segment comprising the parietals, postorbitals, squamosals, and

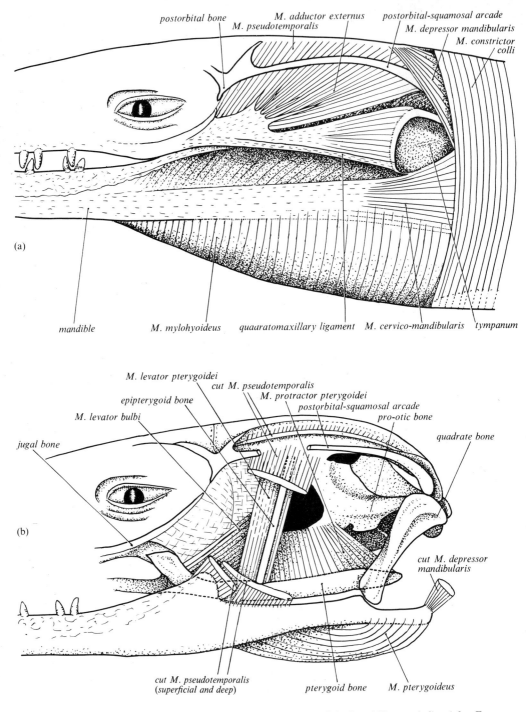

FIG. 42. Diagrams of the jaw muscles of the lizard *Varanus indicus* (after Frazzetta, *J. Morph.* 1962). (a) With skin removed to show the superficial muscles. (b) With the superficial muscles removed or cut to show the deeper muscles.

supratemporals and a maxillary segment composed of the upper jaw bones and the dorsal bones anterior to and including the frontals and post-frontals. The two segments hinge along the fronto-parietal joint and the jugals are separated from the postorbitals, not firmly sutured as in *Lacerta*.

Some other lizards, e.g. the chameleon, have AKINETIC skulls in which there is no movement between the neurocranium and the upper jaws and skull roof. Crocodiles and Chelonians also have akinetic skulls and so does the adult *Sphenodon*. The *constrictor dorsalis* muscle mass of these Reptiles is represented only by the levator bulbi and the *pterygomandibularis* muscle acts as a simple adductor of the lower jaw.

In snakes, kinetism is very well developed but there is a good brain case, which includes dorsal membrane bones such as the frontals and parietals as well as the cartilage bones of the neurocranium; the movement is between this unit and the bones of the upper jaw. This condition is described as PROKINESIS since there is a hinge anterior to the fronto-parietal suture—the exact position of this hinge varies among different types of snakes. Snakes are streptostylic and usually lack a basipterygoid articulation. The *constrictor dorsalis* muscle mass is differentiated into several muscles with different functions and the *pterygomandibularis* is usually large and may have a complex function.

Birds resemble snakes in having a well-developed brain case and usually the upper jaw is movable. There is either a joint between the frontals and the upper beak, a form of PROKINESIS, or the movement is the result of bending somewhere along the flexible dorsal part of the upper beak and this condition is called RHYNCHOKINESIS. Birds are also streptostylic but retain the lower temporal arcade (see p. 188).

Mammals have akinetic skulls and the quadrate no longer takes part in the jaw articulation, which is completely different from that of Reptiles and birds (see p. 192).

In Reptiles, the mouth is opened by the *depressor mandibularis* innervated by the VIIth cranial nerve. This muscle arises from the back of the skull (supraoccipital and sometimes parietal and exoccipital), passes behind the tympanum, and inserts on the posterior ramus of the mandible a short distance behind the articulation between the quadrate and the lower jaw. A *cervicomandibularis* may also be present, arising from the neck and passing antero-ventrally to insert with the depressor mandibularis. The muscles that open and shut the mouth thus both arise from the skull but one group lies posterior and the other anterior to the jaw articulation. A *constrictor colli* may form a sheath round the neck. The levator hyoidei of fishes is represented by the *stapedius* muscle in the middle ear where the hymomandibula bone forms the columella auris (or stapes) conducting vibrations from the tympanic membrane to the inner ear. Transverse ventral muscles form the *mylohyoideus* between the two jaws. The posterior part of this is innervated by cranial VII and the anterior part by cranial V. All the other muscles described in this paragraph are innervated by cranial VII.

The ventral part of the hyoid arch and the remains of the branchial arches form the hyoid apparatus, which plays an important part in swallowing and in movements of the tongue. The *hypo-branchial muscles*, innervated by the XIIth cranial nerve (hypoglossal) are differentiated in relation to these functions and usually include the following: *genioglossus*, from the anterior end of the mandible into the tongue, the protractor of the tongue; *hyoglossus*, from the body of the hyoid into the tongue, the retractor of the tongue; *geniohyoideus*, from the hyoid to the anterior end of the mandible, protractor and levator of the hyoid; *sternohyoideus*, from the dorsal surface of the sternum to the hyoid, retractor of the hyoid which it probably also depresses; *omohyoideus*, from the anterior edge of the scapula to the hyoid, retractor and levator of the hyoid.

Skull of the turtle
(*Chelone midas*)

The skull of the turtle may be taken as an example of a reptilian skull in which the roofing bones form a complete covering without any vacuities. In a posterior view the large foramen magnum can be seen leading into the cavity of the neurocranium. Above this is a roof formed of membrane bones. The skull of the turtle is akinetic and the brain box and upper jaw form a single rigid unit. There is a good joint between the occipital condyles and the long flexible neck.

Bones derived from the cartilaginous neurocranium

The four bones round the foramen magnum, the supra-occipital above, the basioccipital below, and the paired exoccipitals on either side, occupy the usual positions. The supraoccipital is prolonged dorsally and backwards to form a spine. The paired exoccipitals and the basioccipital each contribute equally to the formation of the single condyle. Just outside the exoccipital will be seen the opisthotic bone. Immediately in front of the opisthotic lies the pro-otic bone. These two bones together constitute the auditory capsule.

The floor of the neurocranium ossifies to form the basioccipital behind, with the basisphenoid in front. A triangular portion of the basisphenoid is visible in the ventral view, but anteriorly it is overlapped by the pterygoid bones; there are no basipterygoid processes.

The quadrate formed from the posterior portion of the

palatopterygo-quadrate bar forms an articular facet for the lower jaw. It is a large solid bone, the outer surface being marked by a deep cavity. The posterior surface bears a deep notch through which the hyomandibular or columella auris passes. The pterygoid bones unite with each other in the middle line and cover over the basisphenoid. From the dorsal surface in the middle line, each pterygoid bone sends a process upwards in the postorbital region which meets with a downwardly projecting process of the parietal. Between these two pterygoid processes will be seen the front end of the basisphenoid bone projecting forwards. A membrane bone, the median vomer, lies between the palatine bones. The ventral parts of the palatine bones curve inwards and meet with the expanded lower edge of the vomer, thus forming the posterior boundary of the nares.

The membrane bones of the skull

Neither the upper nor the lower jaws bear teeth but they are covered with horny plates instead.

The premaxillae in front are a pair of small bones that send back processes to meet the palatines and the vomer. The maxillae are large and their ventral surfaces are drawn out to form a sharp edge. Processes from the maxillae extend inwards to meet the palatine and the vomer, thus forming most of the floor of the narial passage.

Uniting the maxilla with the quadrate are two bones, the jugal in front and the quadratojugal behind. The maxilla and the jugal together form the lower border of the orbit.

The nasal bones are fused with the prefrontals. Immediately behind the fused nasals and prefrontals lie the frontals. Behind these again lie the paired parietals. From their ventral surface the parietals send a process downwards in the postorbital region to meet the upward process from the pterygoid. These processes serve as a strut to support the roof of the skull.

The upper border of the orbit is formed by the fused nasal and prefrontal in front and by the large fused postorbital and postfrontal behind. Immediately behind the postfrontal and above the quadrate is the large squamosal. It is marked posteriorly by a large groove which accommodates the *depressor mandibularis* muscle.

Lower jaw

The main portion of the lower jaw consists of the dentary which is fused in the middle line without leaving any trace of the symphysis. The remains of Meckel's cartilage lie in a groove in the inner surface of the dentary. The posterior end of Meckel's cartilage ossifies to form the articular. Posteriorly four paired membrane bones occur. External to the articular and running forwards on the outer side of the jaw is the supra-angular bone. Ventral to the articular and running forwards to meet the dentary, partly on the ventral surface, but mostly on the inner surface, is the angular bone. On the inner side running forwards from the articular bone lies the prearticular bone. The coronoid bone is situated just in front of the supra-angular and prearticular bones on the dorsal surface. It forms a prominent process, to which are attached *adductor mandibulae* muscles used for closing the mouth. The splenial bone is not represented in the turtle.

Beak

No teeth are present on the jaws, but a horny sheath is formed by the skin that covers the biting edges of the premaxilla and maxilla above, and the lower jaw below, forming a beak.

Skull of the soft-shelled turtle *(Trionyx)*

In the skull of the turtle, *Trionyx*, the roof formed by the parietal, squamosal, and postfrontal bones has been emarginated from behind forwards. This exposes the *adductor mandibulae* muscles which are very powerful in this carnivorous animal. Apart from this, the general arrangement of the bones is similar to that of the green turtle and a beak is present.

On the ventral surface, however, the basisphenoid bone is visible for the greater part of its length. The columella auris passes through a foramen in the centre of the quadrate instead of through a notch, as in the turtle. The emargination of the parietal leaves the otic bones exposed, they are hence more easily seen than in *Chelone*.

Skull of the lizard *(Uromastix)*

The skull of the lizard presents a scaffold-like appearance. The brain of the animal is relatively small, and is enclosed in a small cavity at the posterior end of the skull. Identify this cavity in front of the foramen magnum. Its anterior lateral wall is cartilaginous so that the cavity is exposed in front. The origin of the bones either from membrane or cartilage is the same as in the turtle.

This herbivorous lizard probably has an akinetic skull although basipterygoid processes are present. If skulls of other species of lizards are examined, note that the proportions and dentition may be very different. In side view, the postorbital may be separated from the jugal (e.g. *Varanus*) and also from the squamosal (e.g. geckos).

Posterior view

There are the usual four bones round the foramen magnum. The supraoccipital is a flat bone lying at an angle of $45°$ to the opening of the foramen. The exoccipitals on either side have a process from their lower border which, together with the

basioccipital, forms the condyle. The exoccipital has a projection running obliquely outwards. This projection unites on
its anterior face with a similar projection from the bones of the
otic capsule. The bones of the auditory capsule are the
pro-otic and opisthotic bones, but the sutures between these
two are indistinguishable in the lizard.

Dorsal view

Tracing the bones from behind forwards in the middle line
the following are seen. Immediately in front of the supraoccipital lie the fused parietals, which have on either side a
long paired process projecting backwards from the lateral
posterior border. In front of the fused parietals lie the frontals,
also fused to form a single bone. Between the fused frontal
and the parietal in the middle line is the pineal foramen. The
frontal narrows in the middle and is hour-glass shaped. Immediately in front of the frontal lie the paired nasals. In front
of these are the fused premaxillae. The premaxillae bear teeth
and send a process upwards which lies underneath the nasals
and extends backwards as far as the frontal.

Side view

The bones seen in a side view are the premaxilla at the extreme
anterior end and behind this on the ventral border is the
maxilla. The maxilla is a large bone which bears teeth on its
lower border. On its inner border it has a flange which meets
the palatine bone. From its upper border it sends a process
upwards to meet the lachrymal and prefrontal which have
fused to form a single bone. Posteriorly the maxilla articulates
on the outer side with a bone of similar size, the jugal, and on
the inner side with the ectopterygoid (or transpalatine) bone.
The jugal forms the ventral and part of the posterior border
of the orbit. Between the jugal and the parietal at the posterior
border of the orbit lies the small postorbital bone. The
quadrate with which the lower jaw articulates is a flat bone
projecting downwards from a process of the auditory capsule.
Between the upper border of the quadrate and the posterior
end of the jugal lies the V-shaped squamosal. The upper limb
of the V lies close against the parietal process. A small bone
lies closely applied to this part of the squamosal and to the
parietal. This bone is either the supratemporal or the tabular.

Ventral view

Starting from behind and working forwards, the basioccipital
is seen to form a portion of the floor of the brain case. In front
of the basioccipital is the basisphenoid bone. The basisphenoid
has two lateral processes that project obliquely forwards and
abut on the pterygoid bone. Extending forwards in the middle
line from the anterior end of the basisphenoid is a thin needle

of bone, the presphenoid. (This bone is very delicate and is often missing.)

The articular surface of the quadrate is easily distinguished. From the quadrate there extends forwards the pterygoid, a bone replacing part of the palato-pterygo-quadrate bar. The pterygoid is an irregular-shaped bone lying near the quadrate posteriorly and articulating at its anterior end with the palatine bone. Half way along its length it articulates on one side with a process of the basisphenoid and on the outer side with the ectopterygoid bone.

Inside the skull and extending from the upper border of the middle part of the pterygoid is a slender rod of bone, the columella cranii or epipterygoid. The upper end of this bone rests on the lower surface of the parietal. In dried skeletons it usually shrinks away from this position and appears to rest against the auditory capsule. The columella cranii is comparable in function with the broad flange of bone in the Chelonia formed by a long down-growth from the parietal and a short up-growth from the pterygoid. In both cases this bone forms the anterior lateral wall of the brain case and occupies the same position as the alisphenoid in Mammals. The palatine bone extends forwards along the pterygoid and curves upwards in front to meet the downwardly projecting prefrontal. Anterior to this a median flange from the palatine meets, but does not fuse, with its fellow. Between this anterior projection of the palatine and the anterior end of the maxilla lie the paired vomers.

The columella auris is a slender needle of bone lying in a groove on the ventral surface of the lateral processes of the otic bones. It extends outwards passing below the supratemporal (or tabular) through the notch in the upper border of the quadrate to end against the tympanic membrane. The inner end fits into the fenestra ovalis. It serves to convey sound waves from the tympanum to the inner ear. It represents the hyomandibular bone.

Vacuities in the post-temporal region

The upper temporal vacuity is present and lies just behind the orbit. It is bounded above by the parietal, behind by the squamosal, ventrally by the jugal and anteriorly by the postorbital.

Behind the upper temporal vacuity on the posterior face of the skull lies the post-temporal vacuity. It is bounded above by the backward prolongation of the parietal bone, and below by the processes of the exoccipital and opisthotic bones.

Lower jaw

The lower jaw of the lizard consists of a number of bones. The largest bone is the dentary, which bears teeth on its upper border. The dentary is joined with its fellow in the middle line anteriorly by a median symphysis. Posterior to the dentary will be seen a number of small bones. One of these projects upwards to form the prominent process for the attachment of the muscles working the lower jaw. The bone forming this process is the coronoid. Next identify the articular bone which bears the facet for the articulation of the lower jaw with the quadrate. The angular bone lies on the ventral border of the lower jaw. It is a narrow bone that appears on the outside of the jaw at the posterior end and curves inwards anteriorly. Immediately above the posterior end of the angular bone on the lateral border of the lower jaw is the supra-angular bone. The prearticular bone lies mainly on the inner side of the jaw and above the angular, but it extends backwards behind the articular bone to form a prominent process. The splenial is a thin narrow bone which is difficult to leave in place in the preparation of the skeleton. It lies on the inner border of the dentary just in front of the prearticular bone.

In the development of the jaw the articular bone is formed by the ossification from the end of Meckel's cartilage while the remaining bones of the lower jaw are membrane bones.

Teeth

Numerous teeth are borne on the premaxilla, maxilla, and dentary. They are similar in size and form a single row fused to the inner side of the jaw (pleurodont).

Visceral skeleton

The visceral skeleton of the lizard is represented by the reduced hyoid arch. The columella auris represents the hyomandibular bone, and the ventral cartilaginous basilingual plate, supporting the floor of the mouth is formed by the ventral parts of the arch. The basilingual plate has two short cornua in front and two very long ones behind which extend as far back as the arterial arches. The rest of the visceral skeleton has either disappeared or may be represented by the cartilages of the larynx.

Skull of the snake (Python)

The skull of the snake differs from that of the lizard chiefly in the arrangement of the bones that provide the gape of the jaws. The cranial cavity is very long and extends forwards between the eyes, so that there is no interorbital septum. The sides of the skull in this region are formed by the descending processes of the parietal and frontal bones. The premaxillae are fused to form a single median bone which is unconnected, except by ligaments, with the maxilla on either side. (In addition to the maxilla and the premaxilla, both the pterygoid and the palatine bones bear large backwardly projecting teeth; these are fused to the upper or lower edges of the bones (acrodont) and not to the sides of the jaws as in *Uromastix*.) The squamosals are elongated and project backwards from the parietal region of the skull and have sliding articulations with this. The quadrates are freely movable about their articulations with the squamosals and at their lower ends they are connected by loose ligaments to the maxillae and to the pterygoids. The articulations between quadrates and lower jaws allow rotation and lateral movement as well as opening and closing of the mouth. There is no symphysis anteriorly between the two mandibles and there is a joint in each mandible between the dentary (and splenial) anteriorly and the other more posterior bones. These various joints and the absence of a mandibular symphysis make the jaw structure very flexible and allow large animals to be swallowed whole by the snake.

The skull is prokinetic and there is a good brain case which includes the frontals and parietals. The premaxillae and vomer are independently movable. Articulations between the other upper jaw bones allow limited movements but the maxilla, ectopterygoid (or transpalatine), and pterygoid on each side form a functional unit. Each of these upper jaw units can move independently of the other and they are alternately pulled forwards (by the *levator* amd *protractor pterygoidei* muscles) fixed in the prey (by the *pterygomandibularis* muscle) and pulled backwards (also by the *pterygomandibularis*) as the snake pushes its mouth round a large victim. The movements of the quadrate are controlled by a *retractor quadrati* arising from the neck, as well as by the adductors and depressor of the mandible. The *pseudotemporalis* muscle is small but the *adductor externus* is enormous, allowing a powerful bite. The prey is caught and held by the mouth as the coils of the body are wound round it, crushing and killing it by constriction.

In viperine snakes, the skull is short and the short squamosal bones are immovably attached to the parietals. A pair of fangs is borne on the maxillae. The fangs are enlarged teeth with a canal opening at either end. A poison gland (modified parotid salivary gland) opens at the base of the fang and the secretion passes into the canal and so into the wound caused by the fang. (In some poisonous snakes, the fangs have open grooves instead of canals.) Each maxilla pivots on a prefrontal which is itself articulated to the frontal. Forward movement of the pterygoids, produced by the *levator* and *protractor pterygoidei* muscles, is transmitted via the ectopterygoids (transpalatines) to the maxillae causing erection of the fangs. The fangs are retracted when the pterygoids are pulled back, mainly by the action of the *retractor pterygoidei* and *pterygomandibularis* muscles. The quadrates are not involved in the movement of the fangs but must be held firmly by the *protractor quadrati* and *depressor mandibularis* muscles if the *pterygomandibularis* muscle is to produce the movement described. The fangs are erected together with the mouth wide open and they are plunged into the victim by the striking movement of the whole front end of the snake with the head held rigidly on the vertebral column. The poison gland· is squeezed, releasing venom into the wound, by contraction of a superficial part of the *adductor externus* muscle.

The prey is drawn into the mouth, as in constrictors, by alternate movements of the upper jaw units but only the palatal teeth are used. A joint between the pterygoid and ectopterygoid allows the upper jaw to bend so that the maxillary fangs remain retracted while the palatal teeth are pushed forward, pulled down into the prey and then pulled back and released. The active muscles are the *protractor pterygoidei* (pushing the palatine and rest of the upper jaw forwards), the *pterygomandibularis* (pulling the palatal teeth down into the prey and then the upper jaw backwards), the *levator pterygoidei* (pulling the palatal teeth up and out of the prey) and the cycle is then repeated. Synchronized movements of the quadrate result in the lower jaw moving forward under the prey and the actual sequence of muscular movements involved is very complex.

Skull of the crocodile or alligator

The Crocodilia have a complete diapsid skull which is akinetic and monimostylic. The outer surfaces are roughened and covered in life by horny scales. Movable scales protect the tympanic membrane when the animal dives.

The snout is elongated (very markedly in gavials) and the maxillae and nasal bones are especially long. The external nostrils, surrounded by premaxillae and nasals, are close together and face upwards near the tip of the snout. There are single, median frontal and parietal bones, each representing a fused pair, and there is no pineal foramen. The antorbital foramen is lost in modern forms and the prefrontal and lachrymal bones meet the nasal, maxilla, and large jugal on each side. The orbits are small and high on the head. The diapsid vacuities are complete: the upper is surrounded by parietal, postorbital, and squamosal; the lower is surrounded by postorbital, jugal, quadratojugal, and squamosal. The occipital condyle is formed by the basioccipital alone.

A long secondary palate is present separating the nasal passage from the buccal cavity. The premaxillae, maxillae, palatines, and pterygoids all meet midventrally and the internal nostrils are surrounded by the pterygoids and are set very far back immediately above the glottis. Muscular flaps from the palate and tongue can form an efficient valve closing off the respiratory passages from the buccal cavity. The external nostrils can be closed and the crocodile can open its mouth under water without flooding its lungs. There are vacuities below the orbits between the maxilla, palatine, pterygoid, and ectopterygoid on each side. A laterosphenoid (or pleurosphenoid) is present in front of the pro-otic and meets the frontal and parietal dorsally and the basisphenoid ventrally. The opisthotic forms a short paroccipital process on each side and the quadrate is large and hinged to the lower jaw so that no lateral movement can occur.

There are six bones in each lower jaw: the articular (which hinges on to the quadrate), angular, surangular, splenial, coronoid, and large dentary. The two dentaries are firmly fused together anteriorly. There is an external mandibular fenestra between surangular, angular, and dentary.

Teeth

Conical teeth of different lengths are borne in sockets (the thecodont condition) in the premaxillae, maxillae, and dentaries. These are shed and replaced throughout life. The fourth mandibular tooth fits into a notch in the upper jaw between the premaxilla and maxilla in crocodiles, whereas in alligators this tooth fits into a pit in the upper jaw. Gavials have slender jaws with many teeth and the two dentaries are fused for a considerable part of their length.

Muscles
(cf. Fig. 42)

The constrictor dorsalis muscle mass is represented by the *levator bulbi* under the eye. The *adductor externus* and *pseudotemporalis* muscles are simple. The *pterygomandibularis* is large with an extensive origin from the dorsal surfaces of the pterygoid and palatine bones, extending forwards under the eye, and a strong insertion on the posterior border of the lower jaw. This is the principal muscle closing the mouth. The other cranial muscles resemble those of the lizard.

Skull of the bird
(*Meleagris
***gallopavo*, Turkey)**

The skull of birds is characterized by (1) the extensive fusion of the bones with one another so that the sutures become obliterated; (2) the great lightness of the bone, due to its spongy nature and abundance of air spaces; (3) the orbits and temporal vacuities being large and united into one cavity on either side; and (4) the prolongation of the front part of the skull forwards to support the beak which bears no teeth.

The majority of modern birds are prokinetic, with a dorsal hinge between the frontals and nasals. The upper jaw unit includes the upper beak and the palate, the pterygoids, the quadrates, the jugals, and quadratojugals. The quadrates are hinged to the squamosal region of the skull with complicated articular facets and are also articulated with the pterygoids and the quadratojugals as well as bearing the articulations with the lower jaws. Movements of the quadrate lead to raising and lowering of the upper beak with respect to the braincase. The *protractor pterygoidei* muscle (also called *protractor quadrati* or *entotympanicus*), running from the orbital septum to the posterior end of the pterygoid and to the quadrate, contracts and raises the upper beak; it is lowered by the action of the *pterygomandibularis* muscle which arises from the pterygoid and palatine bones and inserts on the posterior angle of the lower jaw and also on the base of the brain case. The basipterygoid processes are usually vestigial but are well developed in Galliformes (e.g. turkey).

In many birds the prokinetic hinge is replaced by a region of bending along the flexible dorsal bar of the upper beak (frontals, nasals, and premaxillae). This condition, rhynchokinesis, is present in the turkey. Except that there is no definite hinge, the mechanical arrangements are similar to those described for prokinetic birds. Most birds including the turkey exhibit a condition called 'coupled kinesis' in which the raising of the upper beak always occurs when the lower jaw is lowered. A strong ligament from the postorbital region of the brain case to the outer side of the mandible is so placed that the mouth cannot open unless the quadrate rotates and so causes raising of the upper beak.

The cranial muscles of birds are basically similar to those of the lizard but differ in relative proportions. The tongue muscles may be very complicated, e.g. woodpeckers, and special muscles may be connected with the syrinx.

Bones derived from the chondrocranium

The four occipital bones form the posterior wall of the skull round the foramen magnum, but the sutures between the supra-occipital and exoccipital bones are obliterated. The basioccipital is smaller than the others and bears a single condyle for articulation with the vertebral column.

The tympanic cavity is an irregular depression in the cranium just behind the angle of the jaw. It is bounded above by the squamosal and below by the basitemporal. The bone lining this cavity is the portion of the bony auditory capsule visible externally on the skull.

The main part of the auditory capsule is formed by the pro-otic bone. The opisthotic bone early becomes united with the exoccipital.

Notice now the position of the cranial cavity in which lies the brain. This cavity is much larger than in Reptiles and lies in the posterior part of the skull. The chondrocranium here, however, is more fully ossified and a bony instead of cartilaginous interorbital septum is present. This bony septum bears a thickened anterior border consisting of the mesethmoid, which is a separate bone in young birds. The remaining part of the septum is thin and is formed by the orbitosphenoid and the presphenoid bones, the latter lying at its lower anterior edge. The lower posterior part of the septum is formed by the thick median basisphenoid which extends backwards to the cranial wall. This lower edge is known as the basisphenoidal rostrum. The pair of laterosphenoid bones form the anterior walls of the brain case on either side and unite with the upper part of the interorbital septum.

The remaining cartilage or partly cartilage bones of the skull are the quadrate, pterygoid, palatine, and articular. The quadrate is a Y-shaped bone. The upper posterior arm articulates by a double head with two facets on the dorsal region of the tympanic cavity, the upper lying on the squamosal and the lower on the pro-otic. The anterior limb or ascending process is directed upwards, inwards, and forwards, while the stem of the Y projects downwards and bears an articular facet for the lower jaw.

The paired pterygoid is a very stout bone situated on the base of the skull, articulating with the quadrate and passing inwards and forwards to articulate with the basisphenoid.

The palatine is a long paired bone articulating with the basisphenoidal rostrum in front of the pterygoid, and passing forwards to unite with the maxillae.

Membrane bones

The membrane bones of the skull are less numerous than in Reptiles. The parietals form the posterior dorsal roof of the brain case and become fused together and to the frontals in front. The frontals are very large and more or less fused in the adult. They form the anterior roof of the brain case and extend forwards forming the upper borders of the orbits. Laterally the frontals are produced into a pair of postfrontal processes behind the orbital-temporal vacuity.

In front of the frontals lie the large nasal bones separated from each other in the middle line by processes from the premaxillae. Laterally the nasals send downwardly projecting processes to meet the maxillae.

In front of the nasals lie the large premaxillae which bear the horny upper beak. The premaxillae are fused anteriorly and each possesses a dorsal 'nasal' process lying between the nasals and a ventral 'palatine' process lying close to the palatine. On either side of the nasal bone an opening, the anterior nares, is left between the two processes of the premaxillae and bounded behind by the nasals. Dorsally the nasals and premaxillae are very loosely attached to the frontals and mesethmoid. Lateral to the nasals lie the paired crescentic lachrymal bones which are loosely united with them.

The maxillae are elongated slender bones lying beside the posterior parts of the palatine processes of the premaxillae. The maxillae are closely associated with the premaxillae in this region, and send maxillopalatine processes inwards on either side above the palatine. The maxilla runs posteriorly along the ventral border of the orbit and is closely connected with the long slender jugal bone. The quadratojugal is similar in form to the jugal with which it is fused in the adult. It is shorter than the jugal and passes backwards to be attached by ligaments to the lower end of the quadrate.

The basal part of the cranium is covered by a large basitemporal bone which lies closely connected with the basioccipital and basisphenoid.

Foramina

The bony skull is pierced by numerous vacuities mainly for the transmission of nerves and blood-vessels. The foramen magnum has already been noticed. Anteriorly the brain case is pierced in the middle line by a large foramen for the transmission of many of the cranial nerves. In front of this foramen the interorbital septum is usually pierced owing to incomplete ossification. The smaller foramina in the skull will be traced in the dog skull.

The single large vacuity in the side of the skull is formed by the fusion of the orbital with the upper and lower temporal vacuities. The post-orbital bar has disappeared and the post-orbital-squamosal bar between the two temporal vacuities is also absent.

Lower jaw

The rami of the lower jaw are fused together in the middle line anteriorly and articulate behind with the quadrate bones. The bones composing the jaw resemble those of Reptiles but the sutures are very difficult to distinguish. The posterior end of Meckel's cartilage ossifies to form the articular bone which bears a facet for articulation with the quadrate. It is drawn out posteriorly into a hook, the posterior articular process.

The remaining bones are ensheathing membrane bones. The dentary is the largest and forms the front and middle portion of the jaw. It bears the horny beak. The supra-angular lies on the posterior outer side of the jaw, and forms the upper part of the mandible in front of the articular. The dorsal surface of the supra-angular is drawn out into the small coronoid process. The angular is the small bone lying along the lower edge of the jaw below the articular and supra-angular. The splenial is a thin bone lying along the inner side of the middle of each mandible covering over Meckel's cartilage.

Hyoid arch

The hyoid arch is composed of a bony rod, the columella auris, derived from the hyomandibular. One end of the columella is expanded and fits into the fenestra ovalis. The rod extends across the tympanic cavity and terminates in a triradiate piece of cartilage attached to the tympanic membrane.

The remaining portion of the hyoid arch is reduced to short cerato-branchials and long basi-branchials, attached to an unpaired plate composed of two portions, often termed the basi- and uro-hyals.

The skull of the dog
(Fig. 44)

In mammals the brain is large and the neurocranium is expanded; the skull is akinetic. There is close association between cartilage and dermal bones making it difficult to distinguish them. A well-developed hard palate separates the nasal passages from the oral cavity. There is only one bone, the dentary, in each lower jaw and the articulation is between this and the squamosal. The quadrate and articular have joined the hyomandibular in the middle ear where they form the three auditory ossicles (see later, p. 195).

There have been considerable changes in the cranial muscles from the typical reptilian condition. The *constrictor dorsalis* muscle group is absent (in akinetic reptiles, such as crocodiles, the levator bulbi is retained but the other muscles of this group are lost). The adductores mandibulae are very well developed and the parts are subdivided as follows.

(1) The *adductor externus* is usually represented by three parts:

(a) the *zygomatico-mandibularis* which arises from the lower edge of the zygomatic arch and inserts on the lower border of the outer face of the dentary; this is often continuous with (b);

(b) the *massetericus* which arises from the maxilla and passes posteroventrally to insert on the lower border of the outer face of the dentary;

(c) the *temporalis* which arises from all round the temporal vacuity, including sagittal crests if present, and inserts on the coronoid process of the dentary.

These muscles vary greatly in relative proportions and shape in mammals with different feeding habits; all three shut the mouth and the last two can cause antero-posterior movements of the lower jaw, used in chewing food.

(2) The *adductor internus pars pseudotemporalis* is represented by two muscles called the *pterygoideus anterior and posterior* (or lateralis and medialis); these arise from the side of the neurocranium and insert on the inner face of the dentary.

If the articulation allows, these muscles can produce lateral movement of the lower jaw, used in chewing.

(3) The *adductor internus pars pterygomandibularis* forms two muscles:

- (a) the *tensor palati*, arising from the pterygoid bones and inserting in the soft palate;
- (b) the *tensor tympani*, arising from the tympanic bulla (usually formed from the angular bone of the lower jaw of Reptiles) and inserted in the tympanic membrane.

This part of the adductor has entirely lost the original functions of closing the mouth and lowering the upper jaw.

The reptilian depressor mandibularis is lost and the mouth is opened by contraction of the *digastricus* muscle which has a double innervation, the anterior belly being innervated by the Vth cranial nerve and the posterior belly by VII. Both bellies are derived from the transverse ventral muscles (*mylohyoideus*) of Reptiles; the right and left sides may be connected by a tendinous fascia or associated with the hyoid plate and the intermediate tendon may be missing so that there is only a single belly. The muscle arises from the styloid process (or from the paroccipital process) and inserts on the lower border of the dentary.

The *constrictor colli* muscle of Reptiles extends forwards external to the jaw muscles and forms the *superficial facial muscles of expression*. There are many slips of these, arising from fascia and inserted into the skin, and they are concerned with movements of the external ear, of the scalp, of the facial vibrissae (long hairs usually with a tactile function), of the eyelids, nose, mouth, lips, and cheek. The details of their arrangement vary greatly with the habits of the animals, see Fig. 43.

As in Reptiles, a *stapedius* muscle is present in the middle ear and there are complex *hypobranchial* muscles concerned with movements of the hyoid plate and tongue. Some of the latter are described under 'Dissection of the Rat'.

Bones derived from the chondrocranium The foramen magnum is surrounded by the usual four bones. The supraoccipital dorsally forms a crest, the sagittal crest, while laterally, where it meets the posterior border of the parietal, it forms the occipital crests. These crests serve for the attachment of muscles used for maintaining the position of the head. The exoccipital bones on either side are drawn out at their edges into two downward-projecting processes that are closely applied to the posterior face of the tympanic bulla. These processes are the paroccipital processes. The inner side of each occipital bone forms a large condyle by which the skull articulates with the vertebral column. The

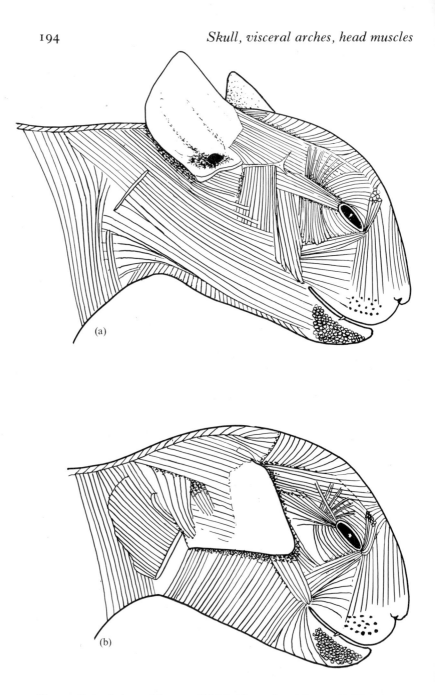

FIG. 43. Lateral views of the superficial facial muscles of the cat. (a) With the ear turned back showing the superficial musculature of the cheek. (b) With the ear turned forwards and some superficial muscles removed to show the deep muscles of the cheek (after Huber, *Q. Rev. Biol.* 1930).

basioccipital is a large flat unpaired bone that extends forwards on the ventral surface as far as the level of the anterior edge of the bulla.

The bones of the auditory capsule differ considerably from those of the Reptile in general appearance. The largest bone in the auditory region is the periotic. The periotic is divided into two portions, the petrous and the mastoid. The petrous portion is formed of very dense hard bone and encloses the auditory organs. It projects into the cranial cavity and it does not appear on the external surface at all. The mastoid portion lies ventral and posterior to the petrous portion. It appears on the outer surface of the skull just external to the exoccipital, and between the exoccipital and the squamosal. The periotic bone ossifies from a variable number of centres but appears to correspond to the pro-otic and opisthotic of other vertebrates. The tympanic bone which forms the auditory bulla forms a symphysis with the ventral surface of the periotic bone. The opening into the tympanic cavity is the external auditory meatus. Across the base of this opening is stretched a membrane, the tympanic membrane (this membrane is almost always destroyed in the preparation of the skull). Abutting against this membrane on its inner surface is a small bone, the malleus. The malleus articulates with another small bone, the incus, which in its turn articulates with the third bone of the chain, the stapes. These three bones together form the auditory ossicles and convey sound disturbances received by the tympanic membrane across the tympanic cavity to the fenestra ovalis, which communicates with the end of the organ of hearing, the cochlea, in the periotic bone.

The origin of the periotic bone has already been mentioned, the remaining bones of the mammalian ear are thought to be derived as follows. The tympanic is formed from the angular bone which grows backwards and becomes cut off from the lower jaw. In many mammals, a cartilage bone, the entotympanic, may surround the middle ear cavity and fuse with the tympanic to form a compound bulla. In some mammals the periotic and bulla fuse with each other and with the squamosal to form a 'temporal' bone. The articular bone of the lower jaw forms the malleus while the quadrate represents the incus. The stapes is pierced by a hole which carries a small artery. From the fact that it is pierced in this way by a foramen for an artery in a similar fashion to the hyomandibular in fish, and from the fact that it occupies the same relative position as the columella auris in Reptiles, the stapes is readily identified as the hyomandibular.

The walls of the cranium are formed partly by the alisphenoid and the orbitosphenoid. The alisphenoid is an irregular-shaped bone that forms the posterior wall of the orbit. Ventrally it sends a plate downwards, the pterygoid plate, which meets the pterygoid bone behind and the palatine in front. The alisphenoid of mammals is homologous with the epipterygoid of Reptiles (see p. 184). The orbitosphenoid forms the anterior wall of the orbital cavity. It articulates behind with the alisphenoid. The articulation is here much obscured by the large foramen lacerum anterius which lies between the two bones. The orbitosphenoid passes forwards and upwards articulating on its dorsal border with the frontal bone. The olfactory capsules of the cartilaginous neurocranium ossify to form the mesethmoid and turbinal bones. The turbinal bones may be seen by looking into the nares from the front end. They consist of a mass of intricately folded lamellae. The internasal septum ossifies behind to form the mesethmoid bone which thus separates the olfactory cavity into two halves. The anterior end of the mesethmoid remains cartilaginous throughout life and forms the septal cartilage of the nose. The posterior end of the olfactory cavity is closed by a screen of bone, the cribriform plate. The cribriform plate is pierced by a series of holes through which pass the olfactory nerves.

The floor of the skull is formed at the posterior end by the basioccipital already described. In front of the basioccipital in the middle line is a single bone, the basisphenoid, and immediately in front of this bone lies the presphenoid. The basisphenoid bears on its dorsal surface a pit, the sella turcica, which accommodates the pituitary body.

Of the palato-pterygo-quadrate bar only the pterygoid and the palatine bones are visible externally, the quadrate having been taken into the ear to form the incus, as already described. The pterygoid bones lie on either side of the presphenoid and anterior portion of the basisphenoid. They form the side wall of the posterior part of the narial passage. The ventral end of the pterygoid bone is drawn out into small backwardly projecting hamular processes. The palatine bones lie in front of the pterygoids and meet together in the middle line anteriorly, where they form the posterior portion of the hard palate.

Membrane bones

The roofing bones of the skull consist of the paired nasals in front. These are narrow bones lying side by side forming the main roof of the olfactory chamber. At their posterior ends the nasals overlap the large paired frontals. The frontals extend downwards laterally and form part of the walls of the orbits meeting the orbitosphenoids. Posteriorly the frontal bones articulate at the sides with the alisphenoid bones and dorsally with the parietals. The parietals are two large paired bones forming the main covering of the posterior portion of the brain. The parietals are separated posteriorly by the anterior projection of the supraoccipital bone. The sutures between the two parietal bones in the middle line are very irregular and small portions of bone are often left as separate bones between the sutures. These bones are known as Wurmian bones.

The membrane bones of the upper jaw are very closely associated with those of the cranium and olfactory capsules. The maxillae are paired bones which send processes upwards to meet the nasals on the dorsal surface. On the ventral surface they bear a palatal process that extends inwards to meet its fellow in the middle line. The incisor teeth are borne on the premaxilla, and the remainder on the maxilla.

The premaxilla is a large bone and like the maxilla it has a nasal process that extends upwards and a palatine process that extends inwards to meet the similar process from the other side. The palatal processes of the maxilla and premaxilla together form the anterior portion of the roof of the hard palate.

The lachrymal bone lies inside the orbit just behind the nasal process of the maxilla. It separates the maxilla from the orbitosphenoid and carries a small foramen, the lachrymal foramen.

The lower border of the orbit is formed by the jugal which articulates at its anterior end with the posterior surface of the maxilla. The jugal meets the forwardly projecting process from the squamosal, the zygomatic process. The arch thus formed is called the zygomatic arch. The squamosal, which has already been identified by its zygomatic process, is the large bone occupying about half of the side wall of the cranial cavity behind the orbit. Although the squamosal occupies this space on the outer wall of the skull, it occupies very little space on the inner surface. It is really a plate of membrane bone overlying the lower portion of the parietal, the posterior portion of the alisphenoid and the periotic bone. It is only by the removal of the squamosal that the proper shape of the alisphenoid bone can be seen.

Vacuities

The orbital cavity is confluent posteriorly with the single temporal vacuity. In some mammals the orbital cavity is closed on all sides by the formation of a post-orbital bar. The temporal vacuity is bounded below by the jugal and squamosal and above by the parietal. The post-orbital bone is absent. This vacuity thus resembles that of the synapsid Reptiles and is the lower temporal vacuity (see p. 176).

The lower jaw

The lower jaw consists of two symmetrical halves united by a symphysis in front. The bone is the dentary of the lower vertebrates. Some of the membrane bones of the posterior portion of the jaw have been incorporated into the auditory region as already described, while the others have disappeared. The upper border of the dentary bears teeth. Behind the teeth it is drawn out into a coronoid process. Immediately behind the coronoid process is the condyle which fits into the glenoid cavity of the squamosal bone. Immediately below the condyle is a small process, the angle. On the inner surface of the dentary just in front of the angle is a foramen for the entry of the V nerve which supplies the teeth. At the anterior end of the outer side is another foramen for the exit of the same nerve.

Foramina of the skull
(Fig. 44)

There are numerous foramina or small holes penetrating the walls of the skull. These holes may actually penetrate the bone or may be small vacuities between the articulations of two bones. They serve to transmit blood-vessels and nerves.

Working from behind forwards these foramina are as follows.

At the posterior end of the skull lies the large *foramen magnum* for the entry of the spinal cord.

Just anterior to the occipital condyles lie a pair of *condylar foramina* in the exoccipital bones on the ventral surface. Each transmits the hypoglossal nerve.

Just anterior to the condylar foramen lies the *foramen lacerum posterius*, a long slit lying immediately behind the tympanic bulla. This foramen transmits the IX, X, and XI nerves, and the jugular and the internal carotid artery.

Just above the lateral posterior part of the tympanic bulla lies the *stylomastoid foramen* which transmits the VII nerve.

On the outer lateral side of the tympanic bulla lies the large *external auditory meatus* leading to the middle ear.

Close to the external auditory meatus just behind the glenoid facet lies the *postglenoid foramen*. This foramen transmits a vein and passes through to the cranial cavity.

On the anterior inner border of the tympanic bulla lie two foramina side by side. The outer of these is the *Eustachian* which communicates with the tympanic cavity. The inner foramen is the *foramen lacerum medium*, which is closed in life by cartilage.

In front of the Eustachian foramen lies the *foramen ovale* in the alisphenoid bone. It transmits the mandibular branch of the V nerve and the middle meningeal artery. Immediately in front of the foramen ovale lies a short canal through the alisphenoid bone, the *alisphenoid canal*, about a quarter of an

inch in length. It transmits the internal maxillary artery, a
branch of the external carotid artery.

Obscured by the ventral wall of the alisphenoid canal lies
the *foramen rotundum*, which transmits the superior maxillary
branch of the V nerve.

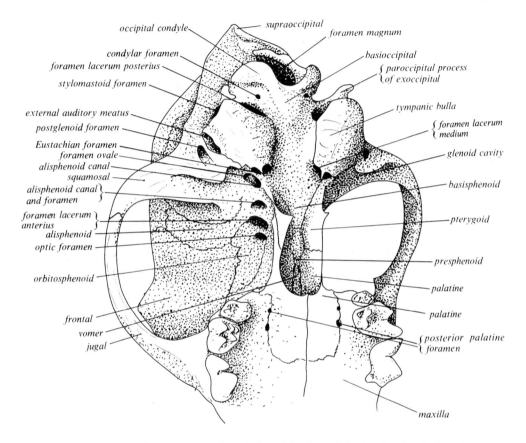

FIG. 44. Ventro-lateral view of the skull of the dog showing the foramina.

In front of the alisphenoid canal lies the large *foramen
lacerum anterius* which transmits the III, IV, and VI nerves
and the ophthalmic branch of the V nerve.

Just in front of the foramen lacerum anterius lies the large
optic foramen, bounded in front by the orbitosphenoid and
behind by the alisphenoid. It transmits the optic nerve.

In the lower anterior part of the orbit just above the posterior molar teeth lie two foramina close together. The lower
of the two emerges through the posterior part of the hard

palate as the *posterior palatine foramen* at the junction of the maxilla and palatine. It transmits the anterior palatine nerve and palatine artery. Smaller posterior palatal openings also communicate with this canal. The upper foramen passes forwards and emerges by the large *anterior palatine foramen* in the hard palate between the canine teeth. It transmits the canals of the organ of Jacobson and the naso-palatine nerves and arteries to the palate. Both these foramina transmit parts of the maxillary nerve.

The *lachrymal foramen* conveys the lachrymal duct through the lachrymal bone and opens into the orbit just behind its anterior border. The lachrymal duct passes forwards into the nasal cavity.

Immediately below the opening of the lachrymal duct into the orbit lies the large *infraorbital foramen* which transmits a branch of the maxillary nerve and emerges through the maxilla on its outer side half way along it. This nerve supplies the sense organs of the mystacial vibrissae.

Teeth

The teeth are embedded in sockets in the upper and lower jaws. They have closed roots and none of them grow from persistent pulps. They are divisible into four groups, incisors, canines, molars, and premolars. The number of teeth in each group is fixed.

The *incisors* are small conical teeth, three in number, situated on either side at the anterior end of the jaws. In the upper jaw they are borne by the premaxilla, all the other teeth being situated on the maxilla. In the lower jaw all the teeth are borne by the dentary.

The *canines* are large pointed teeth and follow immediately behind the incisors, one in each jaw.

The *premolars* and *molars* lie behind the canines. Except for the first premolar they have more than one root. In the adult there are no obvious distinctions in structure between premolars and molars. The distinction arises from the fact that all the teeth in the adult, except the molars, are preceded by temporary milk teeth, which are shed before the animal is full grown, and replaced by the permanent dentition. There are four premolars in each jaw, two molars on the upper jaw and three in the lower. The dental formula thus becomes $i\frac{3}{3} c\frac{1}{1} p\frac{4}{4} m\frac{2}{3}$. The largest tooth in the upper jaw is the fourth premolar. This cuts against the edge of the largest tooth in the lower jaw, the first molar. These two teeth are known as the carnassial teeth and are well developed among many Carnivora, being used for cutting through the bones of the prey.

14 *Skulls of mammals*

General remarks THE bones of the mammalian skull can easily be understood by reference to the description of the dog's skull. In the teeth, however, notable differences occur in the various groups of Mammalia.

Teeth
(Fig. 45)

The teeth of Vertebrates below mammals are usually single-cusped or conodont. The methods of union of the teeth with the jaws are variable. They may be fused to the jaw or attached by ligaments, with or without small intervening bones, or they may be implanted in sockets and then usually have closed roots.

In the mammals the teeth are always implanted in sockets and have closed roots except in rare cases. The number of teeth is definite for any species and never exceeds forty-four altogether, except in cases where secondary reduplication occurs. The teeth are divisible into four types: the incisors, canines, premolars, and molars. There are usually two sets of teeth formed during the life of one animal. The first set is the milk dentition, composed of incisors, canines and premolars. These teeth are subsequently shed and replaced by the permanent dentition composed of incisors, canines, premolars, and molars. The molars represent the most posterior elements of the milk dentition but develop late and are not shed and replaced like the more anterior milk teeth.

In the Mammalia the incisors and canines are always single cusped. The premolars and the molars develop additional cusps which serve to increase the grinding or cutting surface; they usually have at least two roots. A full dentition of forty-four teeth is made up in each jaw of three incisors, one canine, four premolars, and three molars. This is expressed by the dental formula $i\frac{3}{3}$, $c\frac{1}{1}$, $p\frac{4}{4}$, $m\frac{3}{3}$.

There have been many theories about the evolution of the

crown patterns of mammalian molar teeth from single-cusped conodont teeth, but at present none is universally accepted. The names used for the cusps are derived from the Cope–Osborn theory. Unfortunately this theory is almost certainly untrue so that the prefixes used lack the logical basis originally propounded. Each cusp is termed a cone (upper jaw) or conid (lower jaw). One type of primitive tooth, called 'tribosphenic' 'tuberculo-sectorial' or 'tritubercular' consists of an upper molar with three large cusps (trigon): the protocone lingually (on the inner side) and the paracone and metacone labially (on the outer side) with the paracone anterior to the metacone. The corresponding lower molar consists of a trigonid—protoconid labially and paraconid and metaconid lingually, the former anterior—and to this is added posteriorly a talonid (heel) with a hypoconid labially and an entoconid lingually. The protocone occludes with (bites on) the talonid and there may be shearing surfaces between the trigon and trigonid since these parts form an alternating series when the mouth is shut.

The upper molars often have a fourth main cusp, the hypocone, situated posterolingually (the area round it may be called the talon) and the lower molars may lose the paraconid and thus retain only four cusps. Such teeth (upper and lower) are termed 'quadritubercular'. A protoconule may occur between protocone and paracone and a metaconule between metacone and hypocone (or else lingual to the metacone). Accessory ridges of enamel round the margins may form a cingulum—vertical marginal ridges are sometimes present; these are called styles.

The arrangement and the size of the teeth and the crown pattern of the molars is correlated with the diet. Four main types of dentition can be distinguished: (1) the insectivorous type; (2) the omnivorous type; (3) the carnivorous type; (4) the herbivorous type.

Insectivorous and frugivorous forms show the most primitive dentition among modern mammals. Here all the teeth are well developed and the premolars differ from the molars in shape. The molars and premolars have sharp cusps and are not particularly adapted to deal with any special variety of food. In frugivorous forms the molars tend to become flat and are used for crushing fruit.

In omnivorous forms the incisors are of the normal pattern, the canines are well developed and the molars and premolars approach the maximum number of forty-four. The molars and premolars are blunt-cusped, often with secondary cuspules, and the crowns are short. Such teeth are found in the Primates, the bear, and the pig.

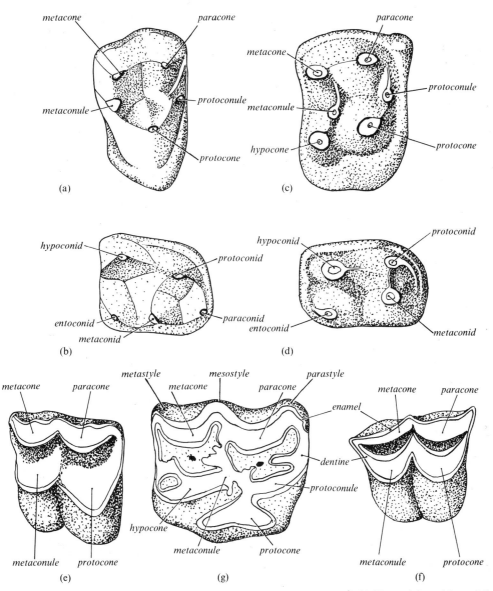

FIG. 45. Molar teeth of placental mammals. (a) and (b) Upper right and lower left molars of *Omomys*, an Eocene tarsioid, with tritubercular pattern. (c) and (d) Upper right and lower left molars of *Hyracotherium*, a lower Eocene horse, showing the evolved quadritubercular pattern. (e) Right upper molar of *Samotherium*, a Pliocene giraffe, and (f) Right upper molar of *Tragocerus*, a Pliocene bovid both showing the selenodont pattern, (g) Right upper molar of *Equus complicatus*, a Pleistocene horse, showing the lophodont pattern. The outer surfaces of the teeth are stippled. The worn exposures of dentine and enamel are white and labelled on (f)–(g) (after Romer. *Vertebrate paleontology*, 1966).

In carnivorous forms the incisors in both jaws form a transverse cutting ridge and the canines are long and well developed. In the upper and lower jaws a pair of large shearing teeth, the carnassials, are developed from the last premolar in the upper jaw and the first molar in the lower jaw. As the carnivorous type of dentition becomes more specialized the carnassials increase in size and the number and size of the molar teeth behind them is decreased.

In herbivorous mammals, the sharp incisors are used for cutting but in some, e.g. Artiodactyls, they are absent from the upper jaw. The canines are reduced or absent and there is a large gap, the diastema, between the incisors and premolars. The molar and premolar teeth tend to become alike in size and crown pattern.

In the more primitive herbivorous forms (e.g. tapir) the molar teeth are low-crowned (brachydont), as they are in most types of detention. In the more specialized herbivorous forms the molar teeth have high crowns (hypsodont) and rise in their sockets as their surfaces wear down (e.g. ox, horse, and rabbit).

In the selenodont tooth (e.g. ox, sheep, koala) the four cusps of the inner and outer border spread out to form crescentic longitudinal ridges (Fig. 45e, f). The valleys between the cusps are filled with cement in the developing teeth before they pierce the gum. As a tooth wears down the different hardness of the cement, enamel, and dentine ensures a rough, uneven surface at all times.

In the lophodont tooth (e.g. kangaroo, rhinoceros, horse) the cusps unite to form ridges that for the most part lie transversely across the tooth. In the development of the teeth of the horse the paracones and metacones are united to form the crescentic longitudinal ridge, while the paraconule unites with the protocone and the hypocone with the metaconule to form two transverse ridges (Fig. 45c, d, g). The deposition of cement and the method of wearing of the tooth is similar to that found in the selenodont type.

In the Proboscidea further transverse ridges are formed on the talon, which is here extremely large.

In the Rodentia, there is a wide variety of tooth patterns based on ridges of enamel which are folded in a complicated manner. The valleys between these ridges are filled by cement.

Monotremata

In the Monotremata the cranium is thin-walled and the sutures between many of the bones are obliterated in the same way as in the bird's skull. The palate extends much farther backwards than in other mammals and the alisphenoid is peculiar. The skull is, however, definitely of the mammalian type as is shown by the presence of ear ossicles, while the dentary is the only bone in the lower jaw and it articulates with the squamosal. Reptilian features in the skull are the presence of a separate postfrontal bone and a post-temporal vacuity.

In *Echidna* the facial portion of the skull (nasals, premaxillae, and maxillae) is extended to form a pointed snout. The pterygoids extend back to form part of the walls of the tympanic cavity. The tympanic bone is a slender ring which is not fused to the skull and does not form a bulla. The stapes is not perforated and resembles the columella of Reptiles in shape. The lower jaw is very slender and there are traces only of a coronoid process and angle. *Echidna* is toothless at all stages of its existence.

In *Ornithorhynchus* the facial portion of the skull is extended forwards to form a flat beak. This beak is supported mainly by the premaxillae which are large and diverge widely at their anterior end. The lower jaw is better developed than in *Echidna*, but the angle and coronoid process are poorly developed. Both the projecting facial beak and the lower jaw are covered in the living animal by a horny integument. The platypus embryo grows a series of calcified teeth most of which are nonfunctional. The dental formula at this stage is $i_{\overline{5}}^{0}$, $c_{\overline{1}}^{1}$, $p_{\overline{2}}^{2}$, $m_{\overline{3}}^{3}$. The first premolar of each jaw is replaced. At 11 weeks old, there remain only two teeth in the upper and three teeth in the lower jaw and these are shed when the animal is half grown and are replaced functionally by horny plates developed from the epidermal lining of the mouth.

Marsupialia

The marsupial skull shows the following primitive features which are characteristic of all primitive mammals: (1) the brain cavity is relatively small; (2) the sutures between the bones are well marked and persistent; (3) the lachrymal bone extends on to the face and its foramen opens either outside the orbit or on its edge, but never within the orbit; (4) the posterior part of the palate is pierced by large vacuities; (5) the periotic bone sends a process backwards which is exposed to a very large extent on the hind surface as the mastoid.

The following features are characteristic of the marsupial skull. (1) The nasals are wider posteriorly than in front. (This feature is seen also in the Perissodactyla and some Artiodactyla.) (2) The premaxillae never, and the maxillae only just, touch the frontal bones. (3) The jugal is large and extends backwards, forming a portion of the glenoid fossa. (*Hyrax* and *Cynocephalus*, which are placental mammals, also show this feature.) (4) The basisphenoid region is relatively elongated. (5) The posterior border of the palate is thickened. (6) The tympanic is small, ring-shaped, and is never fused to the skull. A bulla, when present, is formed by a flange from the alisphenoid bone. This alisphenoid bulla may enclose the tympanic partially, as in *Thylacinus*, or completely, as in *Phascolarctos*. (7) The pterygoid bones are small. (8) The optic foramen and the foramen lacerum anterius are confluent. This, possibly primitive, feature is characteristic of, but not entirely confined to, Marsupials. (9) A small foramen, the post-zygomatic, perforates the base of the zygomatic arch, just below or on the tip of the postglenoid foramen. This feature is distinctive of Marsupials and Monotremes. (10) The internal carotid runs through a foramen in the basisphenoid instead of through the foramen lacerum medium. (11) The angle of the lower jaw is inflected. This, a specialized marsupial feature, is absent only in the degenerate jaws of *Tarsipes* but it is found in some rodents. (12) The condylar foramen is double. (13) The dental formula is usually $i\frac{3-5}{1-4}$, $c\frac{1}{1}$, $p\frac{3}{3}$, $m\frac{4}{4}$. Thus the number of incisors may be higher than in other mammals and the number of molars is greater than the number of premolars. The dentition of the Marsupialia differs from that of the Eutheria in that only one tooth, the last premolar, is replaced. In some forms this tooth remains until the animal is full-grown. In others it is reduced and is absorbed or shed before the permanent molars appear, while in others it has not been detected at all. The anterior teeth are therefore permanent milk teeth.

Superficially, the Marsupial skull bears a strong resemblance to many of the types of skull found in the Eutheria. Thus, the skull of *Thylacinus* bears a very strong resemblance to that of the wolf, while the skull of the wombat is of the rodent type. Not all the Eutherian types, however, are represented in the Marsupialia and the Marsupialia themselves have certain types of skull which are not represented in the Eutheria.

In the Polyprotodontia the number of the upper incisors always exceeds three, while that of the lower incisors is never less than three.

Opossums are omnivorous or insectivorous and their skulls are small. The molar teeth are simple and tritubercular.

In *Sarcophilus* and *Thylacinus* the teeth are tritubercular-sectorial and resemble those of placental carnivores. There are, however, no true carnassial teeth and the canines of the lower jaw fit into a small pit in the upper jaw just in front of the upper canines, a fact which makes them easily distinguishable from similar placental types.

Myrmecobius eats ants and the dentition is specialized. The molars are irregularly cusped and are secondarily increased in number to five in the upper and five to six in the lower jaw.

In the Diprotodontia there is always a pair of enlarged incisors in the lower jaw sometimes with two pairs of small incisors on each side of them. In the upper jaw there are six incisors (three pairs) except in the Phascolomyidae in which there is only one pair. The group is mainly herbivorous and ridged (lophodont) teeth resembling those of the herbivorous Eutheria are developed except in the koala *Phascolarctos*, which has selenodont cusps.

In the kangaroos the milk (last) premolar is persistent for a long time and is molariform. When it is shed the first premolar is shed with it and is not replaced. The last molar does not appear until after the premolars are shed and all the grinding teeth move forward in the skull as the animal ages, in the same manner as in elephants. The lower incisors are long, and, as there is no proper symphysis of the two halves of the lower jaw, these teeth can be used against each other in a scissor-like manner. Owing to the peculiar way in which the teeth are replaced and shed in the kangaroos the dental formula, $i\frac{3}{1}$, $c\frac{0 \text{ or } 1}{1}$, $p\frac{2}{2}$, $m\frac{4}{4}$, never represents the actual condition of the teeth in the skull at any time.

Insectivora

Although specialized in various ways, the modern Insectivora are nearer to the generalized primitive ancestors of the Eutheria than any other form. The skull of the hedgehog (*Erinaceus*) is probably the closest approach to this primitive type living at the present day. It does not show the secondary elongation of the skull so common in mammals of insectivorous habits. The following characters, which may be regarded as primitive, should be noted in the skull of the hedgehog: (1) the postorbital constriction of the skull; (2) the anterior border of the orbit is poorly defined; (3) the lachrymal foramen is on the border of the orbit and the lachrymal bone itself extends on to the face; (4) the posterior border of the palate is thickened; (5) the posterior portion of the palate is perforated by large vacuities; (6) there is a large mastoid exposure on the posterior aspect of the skull; (7) the tympanic bone is small, ring-shaped, and a wing of the alisphenoid forms a bulla, which partly covers the tympanic; (8) the lower jaw has a peculiarly shaped hinder border. Note that characters (1) to (7) also occur in the Marsupialia. In the hedgehog, however, the jugal does not extend backwards into the glenoid fossa, the pterygoid bones are large, the optic foramen is not confluent with the foramen lacerum anterius, nor is the angle of the lower jaw inflected.

The skulls of moles, shrews, Cape golden moles, and tenrecs

agree with that of the hedgehog in the details listed above, except that there may be no postorbital constriction, the zygomatic arch is often incomplete, and the jugal may be small or absent. There is usually no bulla. The nasals are narrowed posteriorly in these animals as in the hedgehog. Tree shrews and elephant shrews (whose relationship to each other is in doubt) have a true tympanic bulla—that of elephant shrews includes the alisphenoid—and the orbit is completely surrounded by a ring of bone. The zygomatic arch is robust and the nasals are expanded posteriorly.

The number of the teeth in the Insectivora may reach the full Eutherian number $i\frac{3}{3}, c\frac{1}{1}, p\frac{4}{4}, m\frac{3}{3}$, as in *Talpa* and *Myogale*, but in other forms the number is reduced. The incisors are usually conical and are set in rows on opposite sides of the jaw. These rows are more or less parallel with one another instead of forming a transverse ridge as in other mammals. This feature is particularly noticeable in the hedgehog and the tenrec, but it is less so in the mole. In the hedgehog and the mole the canines have two roots.

The dentition of the Insectivora generally shows a specialization for a diet of insects. In hedgehogs, shrews, and moles, the molar tooth patterns are 'dilambdodont' with paracone and metacone separate and large—forming W-shaped ridges in moles and shrews—and a hypocone present making a square tooth. Tenrecs and Cape golden moles have 'zalambdodont' molars with the paracone and metacone wholly or partially fused and in the centre of the narrow tooth giving a V-shaped ridge. Tree shrews have dilambdodont teeth but the hypocone may be poorly developed. Elephant shrews eat some vegetation and have large square molars; the last molar is lost and the last premolar is molariform.

Dermoptera (sometimes considered part of the **Insectivora**)

This order contains a single genus *Cynocephalus* ($= Galaeopi-thecus$). The skull has a superficial resemblance to that of the Megachiroptera. There are two incisors in the upper and lower jaws but no upper canines, and there are three premolars and molars in each jaw. The dental formula is $i\frac{2}{2}, c\frac{0}{1}, p\frac{3}{3}, m\frac{3}{3}$. The lower incisors are procumbent, shovel-shaped, and comb-like at the edges. The upper incisors lie at the side of the face, the front of the upper jaw being edentulous, a condition due to the premaxilla extending backwards further than usual. The lower canines and the more anterior upper and lower premolars have prominent sharp-pointed cusps with a number of lesser anterior and posterior cusps in the fore and aft line. The posterior premolars and molars are triangular in plan, multicuspidate, and sharp-pointed. The skull in the adult shows few traces of sutures; these close up early and can only be seen in young skulls. The orbital ring is not complete, but there is a well-marked post-orbital process and a corresponding upward process from the jugal. The tympanic bulla has a well-marked external auditory meatus. The lachrymal bone has a sharp edge on the anterior border of the orbit. The palate has a thickened border but no vacuities. In the lower jaw the ascending ramus is low and the condyle is placed at the level of the teeth. The condyle is transverse and it articulates with a transverse glenoid cavity with a strong revolute process, resembling the Carnivora in this respect.

Edentata
sub-order
Xenarthra

In the Xenarthra the dentition is incomplete or absent. The milk teeth are, as a rule, not functional. Enamel is absent from all the teeth. Incisors are missing altogether, except in *Dasypus*. The zygomatic arch is often incomplete and has a prominent downwardly projecting process from the anterior end.

In *Myrmecophaga* (giant anteater) the facial region of the skull is greatly elongated and produced into a long snout in a manner common among insectivorous Mammals. The premaxillae are small and the long snout is chiefly composed of the maxillae and nasals, together with the mesethmoid and vomer. There are no teeth. The pterygoids meet in the mid-ventral line and form the posterior region of the palate. The zygomatic arch is very incomplete and is without the downwardly projecting process. The lower jaw is very slender and there is no coronoid process.

In the *Bradypodidae* (sloths) there are five pairs of peg-like teeth in the upper jaw and four pairs in the lower. All the teeth are widely spaced apart. The zygomatic arch is incomplete and has a strong downwardly projecting process from the anterior portion. The lachrymal bone is very small and its foramen opens outside the orbit. The premaxillae are very small and the face is very short.

In the *Megatheriidae* (giant sloths) very deeply rooted prismatic teeth, square in cross-section, are present. These teeth are set very closely together. The zygomatic arch is complete and there is a downwardly directed process from the anterior end as in the sloths.

In the *Dasypodidae* (armadillos) the zygomatic arch is complete but there is no downwardly directed process from the anterior portion. The facial region is elongated, but not to the same extent as in *Myrmecophaga*. The teeth are numerous and *Dasypus* is the only edentate possessing incisor teeth.

Pholidota
(= Nomarthra)
and Tubulidentata

In the skull of *Manis* (pangolin) there is a remarkable resemblance to the skulls of the Xenarthran anteaters, but this resemblance is due to convergence and does not indicate any true relationship. In *Manis* the pterygoids do not meet in the middle line as they do in *Myrmecophaga*. There is the usual elongation of the facial region correlated with the insectivorous habits of the animal. There is complete loss of teeth and the zygomatic arch is incomplete, but there is no downwardly projecting process and the jugal is lost.

In *Orycteropus* (aard vark) the zygomatic arch is complete and the mandible is well developed. The adult retains four or five peg-like cheek teeth. These are coated with cement and the dentine is penetrated by many small tubules (forming the basis for the ordinal name).

Carnivora
(Fig. 46)

The modern Carnivora form a natural and well defined order, but absolutely diagnostic characters are difficult to find. It is rather by a combination of characters that the group is defined. In the skull the most marked feature is that the condyle of the lower jaw is transversely placed and is in the form of a half cylinder working in a deep glenoid fossa of a shape that exactly fits it.

The order is divided into two divisions, the Fissipedia and the Pinnipedia. In the Fissipedia, which includes all the terrestrial forms, the digits are free, while in the Pinnipedia, which includes the wholly aquatic forms, such as the seals and walruses, the digits are united by a membrane so that the limb forms a paddle for swimming.

The dentition of the order is never the typical full Eutherian formula of $i\frac{3}{3}$, $c\frac{1}{1}$, $p\frac{4}{4}$, $m\frac{3}{3}$. The full number of three incisors and canines are nearly always present, but a varying reduction, according to the family, takes place in the molars and premolars. In the Fissipedia the fourth upper premolar and the first lower molar are modified to form 'carnassial' teeth. These particular teeth are least modified in the Procyonidae and Ursidae and most highly developed in the Felidae and Hyaenidae. The teeth posterior to the carnassials tend to be reduced or lost in flesh-eaters but retained and used for crushing vegetation in those with mixed diets, e.g. bears and racoons. The upper carnassial consists essentially of a more or less compressed flesh-cutting blade supported on two roots with an inner tubercle on a separate third root. Normally the blade has three cusps, but the anterior one is small and may be lost. The two hinder cusps of the blade with the inner cusp or tubercle constitute the primitive cusps of the original tritubercular tooth, while the small anterior cusp is an accessory structure. The lower carnassial has two roots supporting a crown which consists typically of a compressed trilobed blade with a posterior heel (talonid) and an inner cusp. The inner cusp is the metaconid, the anterior one of the blade is the paraconid, and the posterior of the blade is the protoconid. In the Felidae the blade alone is developed, the heel being vestigial, while in other forms the heel is always present and large.

There is a well marked difference between the skulls of the two groups into which the Fissipedia are divided. These differences centre in the auditory region (see Fig. 46). In the Aeluroidea the bulla and the paroccipital process are closely pressed together, while in the Arctoidea they are separated by an exposure of the mastoid. In the Aeluroidea the tympanic bulla is much dilated; it is also smooth externally, thin-walled, and, except in the Hyaenidae, it is divided into two chambers by a septum. The auditory meatus is short. In the Arctoidea the tympanic bulla is simple and has no septum dividing it into two. The auditory meatus may be long and an alisphenoid canal is present (Fig. 46). Aeluroids lack an alisphenoid canal.

To the Aeluroidea belong the Felidae, Viverridae, and Hyaenidae, while to the Arctoidea belong the Ursidae, Procyonidae, and Mustelidae. The Canidae occupy an intermediate position between the two divisions. In the Canidae the bulla is inflated, but there is only a rudimentary septum; there is also a large exposure of the mastoid behind. The Canidae are usually grouped with the Arctoidea.

In the Felidae the carnassial teeth are strongly developed and there is never more than one much reduced upper and two lower molars. The premolars are also reduced. In the Viverridae there are three or four premolars and one or two molars, the upper molars being sharp-pointed on the inner side. In the Hyaenidae there is only one molar tooth in each jaw together with three premolars.

In the Canidae there are four premolars in each jaw and two molars in the upper and three in the lower jaw. In the Mustelidae there is one molar in the upper jaw, which is expanded on the inner side, and two in the lower. The palate is produced behind the level of the last molar and the bulla is flat; the paroccipital process appears to be pressed against it. The badger is an omnivorous Mustelid and the upper molar is used for crushing. The Ursidae and Procyonidae are omnivorous or herbivorous and their molar teeth have flat tuberculate crowns of the bunodont type and their carnassials no longer shear. In Ursidae there are two molars in the upper jaw and three in the lower. The anterior permanent premolar is reduced and often deciduous. In Procyonidae there are two molars in both upper and lower jaws. The hard palate is produced well behind the level of the last molar as in Mustelidae but the bulla is inflated.

The chief differences in the skulls of the Pinnipedia which

distinguish them from those of the Fissipedia lie in the dentition. In the Pinnipedia there are always less than three incisors in the lower jaw. The cheek teeth consist normally of four premolars and three molars, all more or less alike. These cheek teeth have conical compressed pointed crowns and never more than two roots. Carnassial teeth are absent and the milk teeth are shed very early.

In the Otaridae (eared seals and sea lions) the skull has a postorbital process and an alisphenoid canal. In the Odobaenidae (Trichechidae, the walruses) there is no postorbital process. The skull is very heavy and the dentition is aberrant, the upper canine being very large indeed and forming the tusk used for digging up shellfish on the sea bed. All the other teeth are much reduced. In the Phocidae (seals) the skull has no postorbital process and no alisphenoid canal. The canines are moderately developed. There are five cheek teeth which have a very characteristic appearance approaching the triconodont pattern. Of these five cheek teeth four are premolars, but they can only be distinguished by development as the molars and premolars are all of the same pattern. The skull is usually very light and the ossification is deficient. There is no lachrymal bone or foramen. The brain case is large and the bulla is much swollen. The relations of the bulla, paroccipital process and mastoid ally the seals to the Arctoidea rather than to the Aeluroidea.

Artiodactyla

The skulls of the Artiodactyla may be distinguished from those of the Perissodactyla, to which they converge in some respects, by the absence of the alisphenoid canal and the posterior expansion of the nasals being in many cases slight or non-existent. In the Perissodactyla there is an alisphenoid canal and the nasals are widely expanded posteriorly. In the Artiodactyla the premolars differ in pattern from and are usually simpler than the molars, whereas in the more modern of the Perissodactyla the molars and the premolars are alike.

Living Artiodactyla may be divided into nine families: Suidae, Tayassuidae, Hippopotamidae, Camelidae, Tragulidae, Cervidae, Giraffidae, Antilocapridae, and Bovidae. The first three belong to the suborder Suina, the other six to the suborder Ruminantia (but camels are sometimes put into a separate suborder Tylopoda). In Ruminantia the mastoid is exposed between the exoccipital and squamosal but in Suina the squamosal grows back and conceals the mastoid. In all modern Artiodactyls the face is lengthened and is bent sharply down on the basicranial axis so that the front of the vomer is situated beneath the mesethmoid instead of in front of it as in most skulls. The lachrymal bone extends on to the face.

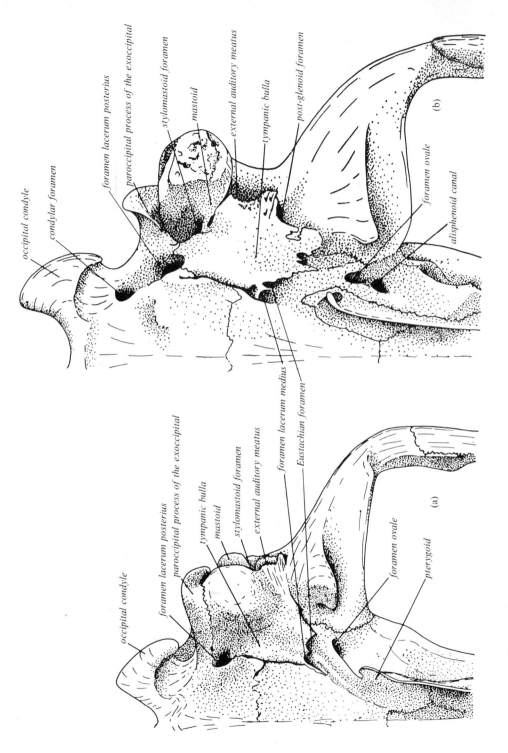

FIG. 46. Ventral views of (a) Hyaena (Aeluroidea) and (b) bear (Arctoidea) skulls showing the tympanic region.

The parietal bone fuses early with its fellow and the frontal bones are large.

Pigs are distinguished by their bunodont, brachydont molars. In the Old World pigs (Suidae) the upper canines form tusks that curve upwards, while in New World pigs or peccaries (Tayassuidae) the upper canines are sharp edged and are directed downwards. In both groups the skull is noticeable for its great length, straight profile and high, back-wardly directed occipital crest. In the wart hog (*Phacochoerus*) the teeth tend to disappear with age, except the canines and the posterior molars. The posterior molars are greatly elon-gated by the enormous development of the talon; this feature, together with the fact that the anterior grinders are worn down and disappear, is somewhat analogous to the condition found in elephants. The stomach in pigs is fairly simple.

In the Hippopotamidae the skull is very large but the brain cavity is relatively small. The incisor teeth are not rooted but grow persistently, forming tusks; those in the lower jaw are procumbent. The canines are very large and curved, and also form tusks. The stomach is three-chambered, but the chambers are all similar and the animals do not ruminate.

In the suborder Ruminantia, the stomach is typically divided into four chambers: the rumen, reticulum, psalterium (or omasum), and abomasum. Food is swallowed without chewing and passes mixed with saliva into the reticulum and rumen where bacterial breakdown of cellulose takes place; acetic, butyric, and propionic acids are produced and ab-sorbed. Later the food is regurgitated into the mouth and masticated thoroughly. It then passes via the reticulum to the psalterium, a chamber with folded walls where fluid is absorbed, and thence into the abomasum where peptic digestion occurs. The molar teeth are selenodont, with four longitudinal crescentic ridges and there is a diastema.

The Camelidae (infraorder Tylopoda) are represented at the present day by the Old World camels and the New World llamas. They are alone among mammals in having oval red blood corpuscles. A postorbital bar is present. One pair of incisors and canines are present in the upper jaws and three pairs of incisors and the canines are present in the lower jaws. Two small premolars are retained in each jaw and the molars are hypsodont. The stomach is three chambered, lacking the psalterium; the rumen is devoid of villi.

The Tragulidae or Chevrotains are a small family with tusk-like canines in the upper jaw and brachydont molars. They ruminate but the psalterium is tubular and without folds.

The Cervidae are characterized by the horns, when present,

taking the form of antlers, which are periodically shed. These antlers are only present in the male, with the single exception of the reindeer, where they are present in both sexes without any distinction of size or form. The antler is a bony outgrowth from the frontal bone and is covered in the growing stage with vascular hairy skin. When fully grown the supply of blood to the skin ceases, the skin then dries and is rubbed or peeled off, leaving the bones uncovered. Each year the antler is shed by the bone just above its base, the 'pedicel', being absorbed. The pedicel remains and by a fresh growth of skin over it a new antler is formed. When the antlers are absent, as in the Moschinae (musk deer), the upper canines form large tusks. In the Cervulinae there are short antlers borne on long bony pedicels. They also have short tusk-like upper canines. In the Cervinae (true deer) antlers are always present and there are seldom any canines. Cervidae have a large antorbital vacuity and two foramina in each lachrymal bone. They are browsers and have brachydont molars. The Moschinae, the most primitive members of the Cervidae, have a gall-bladder, but no other member of the group possesses this structure. All the Cervidae have a ruminant stomach with four chambers.

In the Giraffidae the distinguishing features of the skull are the trilobed, somewhat incisiform canines in the lower jaw, and the short, antler-like horns, which are bony projections from the frontal bone and are present in both sexes in giraffes, but in the male only in the okapi. These horns are covered permanently by the ordinary skin. The molar teeth are brachydont. The Giraffidae are browsers and parallel in the tropics the Cervidae of the temperate zone.

The Bovidae (antelopes, sheep, and oxen), like the Cervidae, have a complete ruminating stomach, but they have a gall-bladder as well. The horns consist of unbranched bony outgrowths from the frontal bones, covered by a layer of horn, which is formed from the epidermis. The horns are never shed and are usually present in both sexes although they may differ considerably in size. These animals graze and the molars are hypsodont. The Antilocapridae (prong bucks) are sometimes grouped with the Bovidae but are, in some respects, intermediate between the Cervidae and Bovidae. In the Antilocapridae the horny covering of the horn is shed annually, but not the bony core. The horn has also a single short branch. The females are hornless.

Perissodactyla The Perissodactyla form a large order in which there are only three living families, the Tapiridae, the Rhinocerotidae, and the Equidae, but the number of extinct forms is very large. In the skull the facial region is lengthened and, in the Equidae, as in the Artiodactyla, is bent sharply downwards on the basicranial axis. The nasals are expanded posteriorly and there is an alisphenoid canal. The cheek teeth form a continuous series, the posterior premolars in all except the earliest forms very closely resembling the molars. The cheek teeth are brachydont in the primitive browsing forms such as the tapir but become extremely hypsodont in recent grazing forms like the horse. A diastema is present. The stomach is simple and there is a large caecum and large colon. There is no gall-bladder.

In the Tapiridae the orbit and the temporal fossa are confluent. The position of the nasal bones, which are very small and elevated, gives the skull a characteristic appearance. The anterior nares are very large and their lateral boundaries are formed by the maxillae instead of the premaxillae, as in the majority of mammals. The living animal has a short flexible proboscis. The nasals are supported by the mesethmoid, which in one species, *Tapirus bairdi*, is completely ossified. The upper canines are reduced and the last incisor is caniniform. The molars and premolars are brachydont and bilophodont.

In the rhinoceros the orbit and the temporal fossa are completely confluent. The nasals are very thick and strong and, extending forwards to the level of the anterior border of the premaxillae, form the supports for the horns, which are composed of fused hairs. As in the tapir, the nasals are supported by the mesethmoid, which is ossified in some extinct forms. The premaxillae are short and, as in the tapir, do not extend backwards to form the lateral borders of the nares. In the Asian rhinoceroses, the first upper incisor and second lower incisor on each side are enlarged and are used to crop vegetation; these tusks are absent in the African species. The cheek teeth are moderately hypsodont and the last upper molar is smaller than the others.

In the horse *Equus* the postorbital bar is complete, the nasals are large and the anterior ends are elevated well above the maxillae, thus leaving a large space for the external nares. There are three pairs of incisor teeth in each jaw and their tips are involuted forming a depression partly filled with cement. As the incisors wear down, the horse's age can be judged by the 'mark' or extent of the pits on the tips of the teeth. Small canines may be present in males but are absent in females. The cheek teeth are extremely hyposodont. The molar tooth pattern has been described on p. 204.

Proboscidea

The Proboscidea living to-day, the African and the Indian elephant, are the survivors of a once numerous series, in which can be traced a progressive shortening of the symphysis of the lower jaw and a corresponding regression of the facial region. In young individuals the skull has a normal appearance, but as the animal gets older the size of the skull is greatly increased (without much increase in weight) by the development of large air spaces in the bones. These air spaces all communicate by narrow ducts with the narial passage. As the air spaces develop the sutures are obliterated. The nasals are very small and the premaxillae, which carry the tusks, are very large and well developed. The nasal passage, directed upwards, is open, as there is no development of maxillo-turbinal bones to close it. In the living animal the external nares are at the end of a long proboscis, and the great development of the surface area of the skull provides attachment for the muscles that work this organ. A striking peculiarity in the skull, which the Proboscidea share with the Sirenia, is the absence of a separate condylar foramen, it being confluent with the foramen lacerum posterius.

The mandible is short with a very high ascending ramus. The symphysis is spout-like.

In the upper jaw there is one pair (the second) of incisors, which form the tusks and grow persistently. There are no canines. The cheek teeth are very peculiar. They are elongated and hypsodont. The grinding surface is marked by transverse ridges with the valleys filled with cement. In the African elephant the transverse ridges are diamond shaped. As these ridges wear down, the dentine of the tooth is disclosed surrounded by a ridge of enamel. In the Indian elephant the ridges are straight. The number of these ridges varies from twenty to thirty. There are altogether six pairs of cheek teeth in both the upper and the lower jaws, three milk molars and three molars. There are usually only two teeth in wear and above the gum at the same time. As the teeth wear down they are pushed forward by those growing behind and eventually drop out, their places being taken by the posterior ones. The milk molars are thus shed and have no premolar successors.

Hyracoidea

The Hyracoidea form a small group of rabbit-like mammals, the skull of which shows certain resemblances to many different groups of animals, e.g. Proboscidea, Perissodactyla, Notoungulata, Lagomorpha, Rodentia, and Artiodactyla. Their actual relationships are still unknown, but a distant one with the Proboscidea is possible. The pattern of the cheek teeth is very like that of the rhinoceros and the premolars resemble the molars as in all Perissodactyla. Primitive features in the skull are the interparietal bones, the alisphenoid canal, and the backward prolongation of the jugal bone. The adaptive features, which give the skull its rabbit-like appearance, are the pair of persistently growing upper incisors, the absence of canines, and the wide diastema between the incisors and the premolars. Features peculiar to the order are that the upper incisors are triangular in cross-section, while the lower incisors have comb-like edges. The postorbital process of the frontal is reinforced behind by a similar process of the parietal, a unique feature among mammals. All the living species are African and are placed in one family, at present including two genera.

Sirenia

The Sirenia are aquatic mammals that live in coastal waters and marshy rivers and crop plants using the fleshy corners of the mouth. The skeleton is remarkably heavy and the skull particularly so. The bones are rough and the sutures are very wide and distinct. At the side of the skull there is a large vacuity between the exoccipital and the squamosal, and in the floor of the skull there is another enormous vacuity formed by the confluence of the foramen lacerum medium with the foramen lacerum anterius. The nasals are very small or absent and the bony narial passage is directed upwards. The premaxillae are large and are bent downwards, their posterior border forming the anterior border of the very large nasal cavity.

There are two living genera, *Manatus* and *Halicore*, while a third, *Rhytina*, only became extinct at the end of the eighteenth century. *Manatus* is remarkable in having an extra large number of molar teeth, which go on increasing indefinitely throughout life, but only about six are ever in wear at the same time. There are two rudimentary incisors in the upper and lower jaws but no canines. In *Halicore* (dugong) there are from five to six molars in each jaw, though they are not always all in wear at the same time. In the upper jaw there is one pair of incisors, which, in the male, are long and grow persistently, forming tusks, while in the female they are small and probably not used. In the lower jaw there are in both sexes four pairs of small conical teeth, which are not used and are covered with a horny pad. These teeth probably represent three pairs of incisors and one pair of canines. In *Rhytina gigas* (Steller's sea cow) teeth are absent altogether and their place is taken by horny pads.

The mandible, like the rest of the skull, is very massive with a high ascending ramus. In *Halicore* the symphysis is bent downwards to conform with the downward projection of the premaxillae.

The group shows some affinities with the Proboscidea: the

absence of a condylar foramen in the skull being a feature
found only in the Proboscidea and the Sirenia.

Cetacea

The Cetacea are profoundly modified for a marine life and are
the most highly specialized of the Mammalia. The skull de-
parts from the usual mammalian type, but it is possible to
trace a connection through the Zeuglodontia with a Creodont
ancestor. The skull is large and the brain case is also large,
although small in comparison with the facial portion. The
supraoccipital is large and the interparietal extends forwards
to the frontal. The parietals are small and fuse with the interparietal. The bones are loosely connected together and the
tympanic is not as a rule ankylosed firmly to the skull, as it is
in most mammals. The tympanic is shell or scroll-shaped
and is formed of much denser bone than the rest of the skull.
When the skull is dried the tympanic readily falls apart. The
tympanic, on account of its hardness, is readily preserved
and these bones are found in considerable numbers wherever
the fossil remains of Cetacea occur. The maxillae are very
large and the premaxillae are long and narrow and extend
back to the external nares which open through the 'blowhole'
on the top of the head.

There are two living suborders, the Mysticeti (whalebone
whales) and the Odontoceti (toothed whales). The Mysticeti
have no teeth and 'whalebone' hangs down from the palate
and filters off zooplankton aided by the tongue. Morphologi-
cally, 'whalebone' is a horny projection from the ridges of the
palate. The skull in Mysticeti is always bilaterally symmetrical. There is a very large supraoccipital which covers
over the parietals posteriorly. The external nares are placed
very far back near the top of the head. The nasals are larger
than in the Odontoceti and they form a roof to the narial
passage. Turbinal bones are present. The ramus of the lower
jaw is arched.

In the Odontoceti the skull has many peculiarities, not the
least of which is the fact that it is not bilaterally symmetrical.
The supraoccipital is very large and forms most of the posterior roof of the skull. The broad interparietal is fused with,
and completely separates, the two parietals. The frontals are
large but are covered by an extension of the maxilla. The nasal
passages are devoid of turbinal bones, the nasal bones themselves are much reduced and take no part in roofing over the
nasal passages. On either side of the nasal passages is a backward extension of the premaxilla, the left of which is shorter
than the right. In front of the nasal opening the face is prolonged to form a narrow rostrum. The rami of the mandibles

are straight. There may be conical peg-like teeth in both the upper and the lower jaws. These teeth are all alike and there is no distinction of form to indicate which are incisors, canines or molars. The teeth may be very numerous in both jaws as in dolphins or in the lower jaw only as in sperm whales. Beaked whales have a single pair of teeth in the lower jaw while the narwhal has one pair in the upper jaw of which the left one is elongated in males to form the long spirally scored 'horn'.

Rodentia
(Fig. 47)

The dentition is the most characteristic feature of the rodent skull. There is one pair of very large incisors in both upper and lower jaws. These incisors have open pulp cavities and grow throughout life. The incisors have no enamel on the posterior face. In many forms the enamel is stained, either black (as in the porcupine) or yellow (as in the beaver). By continuous wear the incisors keep a sharp edge of enamel.

FIG. 47. Lateral outer view of the lower jaw of the porcupine with the bone cut away exposing the persistently growing root of the incisor.

There appear to be only traces of milk incisors. There are never any canine teeth and there is a wide gap or diastema between the incisors and the cheek teeth. Into this diastema the skin can be folded inwards so as to close off the posterior part of the mouth from the incisors when the animal is gnawing, thus preventing chips of bark or wood from getting into the mouth.

The premolars and molars are alike in appearance. Sciuromorphs typically have $\frac{2}{1}$ premolars and $\frac{3}{3}$ molars. Most other rodents have $\frac{0-1}{0-1}$ premolars and $\frac{3}{3}$ molars but *Hydromys* and some others with no premolars have $\frac{2}{2}$ molars and the recently discovered *Mayermys* has no premolars and only one molar in each jaw. The milk dentition is varied and in those forms with a small number of cheek teeth (Muridae and *Hydromys*) there appears to be no milk dentition at all. The permanent cheek teeth vary in shape. They may be hypsodont and rootless, growing throughout life as in the herbivorous forms, or they may be brachydont and rooted as in the omnivorous forms. In the brachydont type, the enamel of the crown is not much folded, and after a certain amount of use this cap of enamel is worn away so that the surface of the tooth comes to consist in old animals of dentine surrounded by a ring of enamel. The brachydont molars of the mice and squirrels show a comparatively simple arrangement of cusps of a more or less bunodont character, but there is a marked tendency in the hypsodont molars of other forms for the development of supplementary cusps. These supplementary cusps coalesce in various ways to form columns and ridges giving different surface patterns. The valleys between the ridges are filled with cement after the ungulate manner. It is difficult to trace any connection between the ridges on the rodent cheek teeth and the primitive trituberculate pattern.

The orbit is not closed behind and the jugal is sometimes prolonged backwards to form a portion of the glenoid fossa. The axis of the glenoid fossa is placed longitudinally and the lower jaw can be moved backwards and forwards. The massetericus muscle is greatly enlarged. In primitive rodents it arises from the zygomatic arch but it is usually divided into two parts one or other of which arises from the maxilla in front of the orbit. In myomorphs, hystricomorphs, and caviomorphs, the medial part passes through the greatly enlarged orbital foramen. The massetericus muscles protract the lower jaw and there may be a posterior or ventral flange for their insertion. The jaw is retracted by the temporalis muscles which also are large. There is no bony symphysis at the front end of the mandibles and these can be separated or pulled together. In the prairie dog *Cynomys* the angle of the lower jaw is inflected as in Marsupials.

The following points in addition to those already mentioned should be noted in the skull. The paroccipital processes of the exoccipital are generally of moderate size, but in the Capybara (the largest living rodent) they are very long. There is sometimes a small interparietal bone. The nasals are large. The premaxilla is large and sends back a process which meets the frontal.

Lagomorpha
(Fig. 48)

The rabbits, hares, and pikas show a remarkable convergence to the Rodentia but are now considered to be a separate order. Their dentition is peculiar because there are two pairs of upper incisors, a small inner pair behind the large anterior pair (see Fig. 48). As in rodents, there are no canines and a long diastema is present. The rabbit has $\frac{3}{2}$ premolars and $\frac{3}{3}$ molars and all these cheek teeth are similar and grow from persistent pulps, as do the incisors.

The maxilla is much fenestrated, giving the skull a characteristic appearance and the infraorbital foramen is small. The axis of the glenoid is placed longitudinally and the lower jaw moves backwards and forwards in gnawing and chewing. The

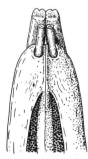

Fig. 48. Ventral view of the premaxillary region of the skull of the rabbit showing the duplicident incisors, one pair being inserted behind the other.

upper tooth rows are further apart than those of the lower jaw (the opposite condition is found in rodents) and the mandibles have a bony symphysis anteriorly. The massetericus and temporalis muscles protract and retract the lower jaw.

Chiroptera

In the Chiroptera the whole skeleton is remarkably light. The skull is the most variable part of the skeleton. There are two divisions, the Megachiroptera (large frugivorous flying foxes) and the Microchiroptera (small insectivorous and occasionally blood-sucking bats). The differences between the skulls of these two divisions are considerable. In both divisions the teeth vary in number, the maximum being $i\frac{2}{3}, c\frac{1}{1}, p\frac{3}{3}, m\frac{3}{3}$. In the Megachiroptera the molar teeth are longer than they are broad and have flat crowns, traversed by a longitudinal groove. In the Microchiroptera the molar teeth have sharp cusps. The crown pattern of the molar teeth serves, in both divisions, to distinguish the skull from that of other mammals. In the Megachiroptera the skull is elongated and the cranial cavity is large; the postorbital processes are long and are traversed by a foramen, except in *Pteropus leucopterus* where the orbital ring is complete; the bony palate is continued behind the last molar and the angle of the mandible is rounded. In the Microchiroptera the skull is shorter and broader than in the Megachiroptera, the postorbital processes are short or absent, the bony palate is not continued behind the last molar, and the angle of the mandible is distinct. The orbits and the nasal openings are large and the premaxillae are short.

Primates

The Primates are characterized by the progressive development of the brain with an accompanying alteration of the facial angles; the preorbital portion of the face becomes progressively reduced and the head bent at right angles to the neck. The lower jaw deepens and its condyle becomes elevated.

The living Primates are divided into four suborders, the Lemuroidea, Tarsioidea, Platyrrhini, and Catarrhini. In the Lemuroidea, the most primitive group, there is a confluence of the orbital and temporal fossae underneath the complete postorbital bar, the lachrymal foramen is exposed on the face and the molar teeth are tritubercular. The upper incisors are spaced apart and may be lost and the two pairs of lower incisors are procumbent and, together with the lower canines (which are incisiform), form a comb used in grooming. In the Lemuridae (true lemurs of Madagascar), there are $\frac{3}{3}$ premolars and the bulla is formed from the periotic bone with the ring-shaped tympanic bone inside it. The Lorisidae (*Galago* and *Loris*) have $\frac{2}{2}$ premolars and the tympanic bone forms a short bony auditory meatus leading into the bulla. The eyes are enlarged and the face is short. The third family of Lemuroids comprises only the aye-aye of Madagascar (*Daubentonia* = *Chiromys*) whose skull is very rodent-like with only one pair of incisors in the upper and lower jaws and no canines.

There is only one living member of the Tarsioidea, *Tarsius spectrum*, notable for its enormous eyes, which look forward, and its reduced nose. The orbits are close together and the alisphenoid is expanded so that there remains only a narrow orbital fissure separating the orbit from the temporal fossa. The upper incisors are close together, the lower incisors are vertical and the canines are caniniform. The bulla and tympanic bone resemble those of Lorisids.

In the Platyrrhini and Catarrhini, which are sometimes grouped together as Anthropoidea, the orbit is completely closed and separated from the temporal fossa, the lachrymal foramen is within the orbit and the upper incisors are close together. The cranial cavity is expanded and the foramen magnum faces downwards instead of backwards. The two mandibles have a firm bony symphysis.

The Platyrrhini (New World monkeys) have in the skull a broad cartilaginous internasal septum and a large bulla enclosing a tympanic ring (as in lemurs). There are usually three premolars and three molars but marmosets (Callithricidae) have only two molars in each jaw and they also have procumbent incisors and incisiform canines as in lemurs.

In the Catarrhini (Old World forms) there is a narrow internasal septum and a bony tube-like auditory meatus leading to the tympanic cavity; there is no bulla. There are never more than two premolars. The molar pattern is bilophodont in the herbivorous Cercopithecidae (Old World monkeys) some of which have long muzzles, e.g. macaques and baboons. In the family Pongidae (apes) the molar teeth are bunodont with a well-marked trigon and small hypocone in the upper teeth and the hyoconulid well developed to give five cusps in the lower molars. The molars form two parallel rows behind the enlarged canines. Supraorbital crests are large in the chimpanzee and gorilla. In the Hominidae, the brain case is larger and the tooth row is rounded with the canines no larger than the incisors. The premaxilla cannot be distinguished on the face and there is a 'chin' on the lower jaw. Supraorbital ridges, prominent in Neanderthal man, are reduced in modern man, *Homo sapiens*.

15 Vertebral column and ribs

The dogfish and skate

THE primitive skeletal rod which runs along the longitudinal axis of the body of the Vertebrate is the notochord. This lies immediately below the spinal cord. In *Petromyzon* the notochord stretches from the infundibulum of the brain to the extreme posterior end of the body. The notochord is an elastic rod of uniform thickness throughout except at the two ends, where it tapers to a point. It is surrounded by two sheaths, the elastica interna and the elastica externa.

The vertebral column is formed in one of two ways, either by the calcification of the notochord itself or by the formation of paired cartilages, and subsequently bones, in the connective tissue surrounding the notochord. In *Petromyzon* the rudiments of the vertebral column are present in the form of small paired cartilages arising on either side of the nerve cord. There is a pair of cartilages in each segment.

In the dogfish or skate the vertebral column is made up of vertebrae. Each vertebra is situated between two myotomes so that it is intersegmental in position and the muscles of two segments are attached to it. The muscle fibres run between adjacent myosepta in the longitudinal plane of the body and can thus bend the intervertebral 'joint'. These joints are not very flexible in Elasmobranchs. Movement is limited to the horizontal plane. Each vertebra is composed of a neural arch which surrounds the spinal cord and a centrum or body of the vertebra which partly replaces the notochord. There are accessory neural elements between the neural arches so that there is a continuous cartilaginous sheath for the spinal cord. In the posterior region the haemal arches form a canal for the caudal vein and artery and there are accessory haemal elements present. The vertebral centra are biconcave (amphicoelous) and the spaces in between the centra are filled with notochordal tissue. The roots of each spinal nerve emerge separately from either side of the neural arch, the dorsal root from the anterior and the ventral root from the posterior side.

In the anterior part of the vertebral column the centra are drawn out into short horizontal transverse processes, which bear at their ends movably articulated cartilaginous ribs. These appear to be homologous with the dorsal ribs of Teleosts and with the ribs of Tetrapods. The haemal arches form a continuous series with the ribs. In the skate there is a specialization of the anterior portion of the vertebral column. The vertebrae are fused together to form the large vertebral plate between the pectoral girdle and the skull.

In the dogfish the tail is heterocercal with the vertebral column bent upwards posteriorly. The hypochordal lobe of the tail is of greater area than the epichordal lobe which lies dorsal to the vertebral column. The haemal arches are prolonged into haemal spines supporting the ventral hypochordal lobe and short neural spines support the epichordal lobe. Lateral movement of the heterocercal tail propels the body forwards and also generates forces lifting the tail. The fairly rigid pectoral fins (see later) are held horizontally and, together with the flattened ventral surface of the head, generate lift at the anterior end of the body. The median fins allow control of rolling and yawing movements of the body. In the dogfish there are two dorsal and one anal (ventral) fins. These fins are supported by rods of cartilage, the radials, extending into the web. Overlapping the radials and supporting the outer parts of the web are horny fin-rays called ceratotrichia.

In the skate the tail is straight and the caudal lobes are small as are also the two median dorsal fins. There is no anal fin. The tail functions as a rudder and the body is propelled by undulations of the greatly expanded pectoral fins (see p. 241).

The whiting or cod or trout

The vertebral column of the Teleost fishes is formed of bone. The centrum is deeply concave both anteriorly and posteriorly (amphicoelous) with the spaces between adjacent centra filled with notochordal tissue as in the dogfish or skate. The neural arches arise from the dorsal surface of the centrum and project backwards uniting to form the neural spine. The neural arches at the sides do not completely invest the nerve cord as in the dogfish. Projecting forwards from the base of the neural arch on either side is a small process, the anterior zygapophysis, which articulates with the process directed backwards from the dorsal surface of the centrum of the preceding vertebra, the posterior zygapophysis. It should be noted that the posterior zygapophysis fits outside the anterior zygapophysis and not inside as in higher forms. The intervertebral joints allow some lateral flexion and this is produced by the contraction of the myotomic muscles, divided into dorsal epiaxial and ventral hypaxial muscles.

The transverse processes are large projections of bone from the lower lateral borders of the centrum. In the tail region these processes meet together in the middle line to form the haemal arches.

In the trunk region the transverse processes have at their ends movably articulated ribs, two to each transverse process. The upper or dorsal ribs pass into the horizontal septum which separates the myotomes into the dorsal and ventral portions, and are comparable to the ribs of Elasmobranchs and to those of higher forms. The ventral ribs pass just inside the peritoneal lining of the body cavity.

In the anterior region of the vertebral column the dorsal ribs have shifted their articulation from the transverse process to the upper part of the ventral rib. In the most anterior vertebra, the atlas, the centrum is considerably narrowed below and widened above. The spine projects upwards almost vertically and not backwards as in other vertebrae behind. The anterior zygapophyses are large and come into intimate union with the posterior part of the skull. As this attachment of the atlas to the skull is firmer than the attachment of the following vertebrae to the atlas, the atlas is often left attached to the skull in the preparation of the skeleton.

In most adult Teleosts the tail is symmetrical externally but asymmetrical internally and is described as homocercal. In the trout the last three centra are inclined slightly upwards and there is a short urostyle surrounding the remnant of the notochord. There are five hypurals (haemal spines fused with radials) below the axis and the bony fin rays or lepidotrichia articulate with these. In the cod and whiting the tail is symmetrical internally and is termed gephyrocercal. All the vertebrae are horizontal and there is no urostyle but a terminal hypural. The detailed structure of the tail varies within the Teleostei and so does the arrangement of the specialized caudal muscles.

Lateral movement of the externally symmetrical homocercal tail propels the body forwards but does not generate lift. The pectoral fins of Teleosts (see later) are usually mobile and can be used as rudders or brakes or for propulsion. The median fins of Teleosts vary in number and size but in all the web is supported by lepidotrichia. These are forked proximally and the ends fit over the distal radial bones. There are usually proximal, median, and distal radial bones. The proximal bones, sometimes called basals or supraneurals (or infrahaemals) are long and their inner ends lie deep in the body and may make contact with the neural (or haemal) spines of the vertebrae. Primitively they were more numerous than the vertebrae. The median and distal radials are small. The radial muscles, which move the fin-rays, arise from the proximal elements and insert on to the base of the lepidotrichia. Many Teleosts use undulations of the median fins in swimming and can pass such undulations anteriorly or posteriorly along the length of the fin. It is characteristic of more advanced Teleosts, e.g. the perch, that some of the median fin rays (as also some of those of the paired fins) should be spiny. All the fin-rays of the trout, cod, and whiting are soft.

Many primitive Teleosts such as the trout have intermuscular bones developed in the myosepta usually of the epiaxial muscles. These are not found in more advanced Teleosts such as the cod and perch.

The salamander   In the newt or salamander the vertebrae are more elongated than in the cod and the notochord is much more constricted. The centra are amphicoelous.

The neural spine is inconspicuous and the neural arch completely invests the spinal cord. The vertebrae articulate with each other by anterior and posterior zygapophyses in the normal tetrapod fashion, viz. the articular facets of the anterior zygapophyses face inwards and upwards while the facets of the posterior zygapophyses face downwards and outwards (cf. whiting). The trunk and neck vertebrae have transverse processes with which the short ribs articulate. In the tail region haemal arches are present. The first vertebra is modified in connection with its articulation with the two condyles of the skull. It has no transverse processes.

In the region of the pelvic girdle two or three vertebrae are slightly enlarged in connection with the attachment of the muscles of the pelvic girdle. The ilia are attached to the modified rib on one of these vertebrae, which thus forms the sacrum.

The facets between the zygapophyses are horizontal and allow considerable lateral flexion but very limited dorso-ventral flexion or rotation about the long axis. When the salamander walks or swims, the body undergoes considerable lateral undulation and this is brought about by the mass of epiaxial muscles extending from the back of the skull to the tip of the tail. Hypaxial muscles in the tail resemble those of fishes and assist in swimming but in the trunk region they are differentiated into the rectus abdominis (the ventral longitudinal muscles) and the obliquus abdominis forming lateral sheaths to the visceral cavity.

In the Anura the vertebral column consists of nine vertebrae only and the tail is represented by an unjointed bony urostyle.

The first vertebra bears two anterior facets for the occipital condyles of the skull. There are seven procoelous trunk vertebrae with short transverse processes and a single sacral vertebra with strong transverse processes articulating with the ilia. The body muscles are essentially like those of Urodeles but there are no caudal muscles.

The lizard

In the vertebrae of the lizard the centra are concave in front and convex behind, a condition known as procoelous. The transverse processes of the neck and trunk region are short, but in the tail region they are long. Except for the first two, the vertebrae in the neck and trunk region in front of the pelvic girdle bear movably articulated ribs. In the Reptile, as in the bird and mammal, the first two vertebrae of the column are modified for the attachment of the skull. The atlas or first vertebra is a ring of bone. Its centrum has become detached and is attached to the front of the next vertebra, the axis, forming the odontoid process. Between the sides of the atlas lies the transverse horizontal ligament that takes the place of the bone lying below the spinal cord. Intercentra or subvertebral wedge-shaped bones are present in the cervical region.

The sacral vertebrae are two in number with shorter centra than the rest. They bear large expanded transverse processes directed outwards, to which are attached the ilia of the pelvic girdle.

Anteriorly the caudal vertebrae are large and well developed, with transverse processes. They bear Y-shaped chevron bones attached between the centra.

The ribs consist of a dorsal bony portion and in the anterior thoracic region of a ventral cartilaginous portion. The ribs articulate with the centra of the vertebrae by a single head. The cervical ribs are short and stout. The first five thoracic ribs are round and attached ventrally to the sternum. The more posterior ribs, or 'floating' ribs, do not reach the sternum.

The zygapophyseal facets are horizontal, as in the salamander, and the vertebral column has very similar mobility. The joints between the centra are stronger because the vertebrae are procoelous. The epiaxial musculature is differentiated into three longitudinal series which are well marked in the trunk region and the lizard undulates its body when it walks. The hypaxial musculature is also well differentiated and the obliquus is divided into several sheets. Ventilation of the lungs is produced by movements of the ribs using the iliocostalis (part of epiaxial) and obliquus muscles. The musculature in the tail is quite fish-like but the caudofemoralis muscles arise below the anterior caudal vertebrae (and see p. 247).

In some Reptiles, e.g. *Sphenodon* and the crocodile, 'abdominal ribs' are present in the ventral abdominal wall as well as true ribs. Abdominal ribs are simply ossifications in the tendons of the rectus abdominis muscle, and are not cartilage bones as are true ribs.

The snake

In the snake the vertebrae are numerous and similar in form apart from the atlas and axis. The centra are large and procoelous, bearing a large round facet on the posterior end for articulation with a concave surface on the succeeding vertebra. The neural spines are well developed. The anterior and posterior zygapophyses are normal in shape but between them, at each end of the vertebra, another pair of articular facets is formed. On the anterior end of the neural arch a pair of facets faces downwards and outwards opposite and between the anterior zygapophyses. These are the zygosphenes. The zygosphenes fit in a hollow of the posterior side of the neural arch of the preceding vertebra. In this hollow a pair of facets is present on either side which face inwards and upwards and are known as the zygantra.

At the anterior end of the centrum lie the short transverse processes with which the single-headed ribs articulate. Ribs are well developed along the whole length of the trunk, which occupies most of the body. Ventilation of the lungs is produced, as in lizards, by movements of the ribs.

The complicated vertebral joints make the column very strong, reducing the risk of dislocation. As in the lizard, vertical flexion and rotation are very much limited but considerable lateral flexion is possible. The 'serpentine movement' of the majority of snakes is the result of lateral flexion, the sides of the body being pressed against objects in the environment and the body then being levered forwards. Some snakes, e.g. vipers, move with the body held straight and the ribs held rigid but the ventral scales are protracted and retracted using hypaxial muscles running between these scales and the ribs. The epiaxial muscles of snakes are very tendinous and have complex interconnections.

The turtle or tortoise

The Chelonia are unique among living vertebrates in having dermal armour forming a box into which the head, limbs, and tail can be withdrawn. The structure of the armour is described later (p. 247) with that of the girdles and limbs. The total number of vertebrae is smaller than in other orders of Reptiles and the column is clearly differentiated into neck, trunk, sacral, and caudal regions.

There are eight cervical vertebrae and the first two form an atlas and axis as in the lizard. The neck is always very mobile but the pattern differs in the three main types. In the majority of Chelonia (Cryptodira), the neck is bent in the vertical plane. The centre of movement is at the level of the fourth or fifth cervical vertebra which has a biconvex centrum. The vertebrae anterior to this are opisthocoelous and those posterior to it are procoelous and the first trunk vertebra has a socket anteriorly. In *Trionyx* (probably an aberrant Cryptodire) the biconvex vertebra is the seventh or eighth and all those anterior to it are opisthocoelous. In Pleurodira, which bend the neck in the horizontal plane, the second or third cervical vertebra is biconvex and all those posterior to it are procoelous. The zygapophyseal articulations are complex and the whole neck is strong but flexible.

The ten trunk vertebrae are immovably joined together and to the carapace. There are two sacral vertebrae, with whose short ribs the ilia articulate, and a small number of procoelous caudal vertebrae.

The trunk muscles are very much reduced except for the obliquus muscles which cause the flank membrane movements used for ventilation of the lungs. The neck muscles, however, are well differentiated and there are typical reptilian muscles in the tail. The armour is used as origin for neck and tail muscles.

The crocodile   The vertebral column of the crocodile resembles that of the lizard in most respects and the musculature is very similar. There is a small median pro-atlas between the atlas and the occipital region of the skull and the axis bears an odontoid process. There are about twenty-four presacral vertebrae usually divided into nine cervical, eleven dorsal, and four lumbar. The centra are procoelous and the zygapophyseal facets are all horizontal. The first sacral vertebra is coeloplatyan and the second is platycoelous and the two are firmly bound together. The first caudal vertebra has a biconvex centrum indicating a centre of movement there and the thirty or more posterior caudal vertebrae are procoelous with chevron bones below them. The crocodile swims by lateral movements of its tail. When moving rapidly on land, it raises the body and holds the limbs under the body.

The ribs are double-headed. In the cervical region the capitulum articulates with the centrum and the tuberculum with the transverse process, but in the trunk both processes articulate with the transverse process of the neural arch. The lumbar vertebrae are without ribs. The caudal ribs are fused with the vertebrae. The thoracic ribs bear uncinate processes.

Seven pairs of abdominal ribs (gastralia) are present ventrally. These are ossifications in the tendons of the rectus abdominis muscles.

The Pigeon

In the bird the vertebral column is distinctly divided into regions and each region has a more or less definite number of vertebrae. The atlas and axis are of a similar type to that found in Reptiles. These two are followed by twelve others and to-gether, in the pigeon, they form the cervical region. These vertebrae are large and articulate freely with each other. The articulating surface of the centra is of a peculiar shape, usually described as saddle-shaped. The centra have short transverse processes from the ventral anterior border projecting out-wards, to which are fused the reduced backwardly projecting ribs. Only in the posterior region are the ribs free. The rib is fused both to the centrum and to the transverse process, and thus a canal (vertebrarterial) is formed enclosed by bone through which the vertebral artery passes. The number of the neck vertebrae is fixed only for the species and varies from eight to twenty-three.

The five thoracic vertebrae are distinguished by the fact that their ribs meet the sternum ventrally. The first four thoracic vertebrae are fused together and the last thoracic is fused to the next posterior vertebra, the first lumbar. The dorsal and ventral parts of the ribs are both ossified and are movably articulated with one another laterally. The rib articu-lates with the vertebral column by two processes. The head or capitulum articulates with the centrum and the tuberculum articulates with the small transverse process. Between the capitulum and tuberculum lies the continuation of the verte-brarterial canal. The last cervical and first four thoracic ribs bear backwardly directed uncinate processes from the middle of their upper portions. These processes overlap the ribs be-hind them. There are six lumbar vertebrae which are fused together and fused also in front with the last thoracic and behind with the sacral vertebrae, while the sacral vertebrae are in turn fused with the first five caudal vertebrae. These fused vertebrae are united or fused with the pelvic girdle. There are six free caudal vertebrae followed by a flat bone, called the pygostyle, which consists of four vertebrae fused together.

The cervical region of the bird is very mobile but the joints are strong. Most of the rest of the vertebral column is rigid and the reduced caudal region supports the fan of tail feathers. The musculature of the neck is very complex. That of the trunk is reduced except for those muscles taking part in the respiratory movements of the ribs and sternum. The tail muscles are concerned with raising, lowering, spreading, and folding the tail fan, movements that are important during flight.

The Mammalia The vertebral column of the Mammalia is divided into distinct regions and there is a tendency for the number of vertebrae composing the region to be constant in each order. A characteristic of the vertebrae of the Mammalia is the presence of epiphyses. These are disks of bone which are fused to the centrum on its anterior and posterior faces in the adult animal. The articulating surface of the centrum is flat. The regions of the vertebral column are five, viz. cervical, thoracic, lumbar, sacral, and caudal.

Cervical region The cervical vertebrae always number seven, except in *Bradypus*, the two-toed sloth, and *Manatus*, which have only six, and *Choloepus*, the three-toed sloth, which has nine. Lengthening or shortening of the neck is effected by the elongation or shortening of the individual vertebrae and not by increase of the number. The first cervical vertebra, the atlas, is a bony ring with wide, wing-shaped transverse processes. The second cervical vertebra, the axis, has a process, the odontoid process, attached to the anterior face of the centrum. This process is formed by the centrum of the atlas fusing with the centrum of the axis. The transverse processes of the first six cervical vertebrae are pierced by a canal, the vertebrarterial canal,[1] which serves as an easy means of distinguishing these cervical vertebrae from those of other regions. The vertebrarterial canal is formed as in the Bird, by the head of the rib fusing with the transverse process, the canal representing the space between the tuberculum and capitulum of the rib. It follows, therefore, that movable ribs do not occur in the cervical region. They may, however, be seen in certain cases in the posterior cervical vertebrae, as in Sirenia and the brown bear. They are, however, very small and never reach the sternum. The vertebrarterial canal is absent from the last cervical vertebra (the ninth) in *Choloepus*.

In Monotremes the cervical vertebrae, except for the atlas of *Echidna*, have no zygapophyses. In the Marsupials the lower arch of the atlas is often incomplete. Fusion of the cervical vertebrae occurs in certain animals. In *Tatusia*, not only the centra, but also the neural arches, are fused together to form a solid tube. In the jerboas all the cervical vertebrae

---

[1] The transverse processes of all seven cervical vertebrae are occasionally pierced by the vertebrarterial canal. It happens commonly in the Marsupialia and is characteristic of them, although by no means all of them have this character. It occurs also in some rodents, in *Hyrax*, in *Cynocephalus*, and also in the hippopotamus.

except the atlas are fused together. In the Cetacea the vertebrae are compressed and in certain forms, e.g. right whale and grampus, they are fused together. There is no odontoid process in Cetacea, except in the rorqual, where one is just distinguishable, and the vertebrarterial canals are incomplete. Incomplete vertebrarterial canals occur also in the African ,elephant. The articulating surfaces of the centra of all cervical vertebrae behind the axis are usually flat or amphiplatyan, but in the ungulates they are opisthocoelous, more so in the Perissodactyla than in the Artiodactyla.

The neck is mobile in most mammals but movement is strictly limited in those where there is fusion of cervical vertebrae. The joints between the atlas and the occipital condyles permit dorso-ventral movement. The joint between atlas and axis allows rotation and the other cervical joints allow lateral flexion with some dorsoventral movement and some rotation. The muscles between neck and head are well developed as well as the cervical muscles.

Thoracic and lumbar regions

The number of the trunk vertebrae is by no means fixed. Excluding the sacral vertebrae, the maximum number is thirty, which occurs in *Hyrax*, where twenty-two of these are thoracic and bear ribs, and eight are lumbar. The most usual number is from eighteen to twenty-three. In certain orders the number is fixed, while in others it varies considerably.

The thoracic vertebrae are distinguished from the lumbar by bearing ribs, while the lumbar do not. The ribs are curved rods which are movably articulated at their dorsal ends with the vertebrae. The dorsal end of each rib is rounded, forming the head or capitulum. The capitulum fits into a depression formed partly by the vertebra corresponding to the rib, and partly by the following vertebra. Just behind the capitulum there is a rounded prominence, the tuberculum, which articulates with the transverse process of the vertebra. The dorsal portion of the rib is ossified but the sternal part is formed of cartilage, which is in many cases calcified or imperfectly ossified. The ribs of the Mammalia never have uncinate processes. The anterior ribs are united at their distal ends with the sternum and those that are so united are called 'fixed' ribs, the posterior ones that are not so united are called 'free' or 'floating' ribs. The thoracic vertebrae have a high neural spine which forms the attachment for the muscles supporting the head and neck. The lumbar vertebrae also possess spines, but these are usually lower than those of the thoracic vertebrae. The spines of the anterior thoracic vertebrae usually point backwards, and those of the posterior thoracic and lumbar forwards, so that all the spines converge to a point which is called 'the centre of motion'. This feature is especially well marked in the Ungulata. In some animals, e.g. rhinoceros, sloths, sea lions, all the neural spines, both thoracic and lumbar, point in the same direction, backwards.

The lumbar vertebrae are larger and longer than the rest. The transverse processes are usually large and forwardly projecting. From the ventral side of the centrum of some of the lumbar vertebrae a spine projects downwards known as the hypapophysis. The neural arches are produced anteriorly to form processes projecting above the anterior zygapophyses, known as the metapophyses, which stand on either side of the neural spine. Posteriorly between the centrum and the posterior zygapophysis, the neural arch may be prolonged laterally as the anapophyses. In some forms extra articulating facets are formed between successive transverse processes.

The zygapophyses of the anterior thoracic vertebrae have horizontal articulating facets and these joints allow considerable rotation and lateral flexion but very little dorsoventral movement. The three longitudinal series of epiaxial muscles are well differentiated and the median ones are very complicated with many tendons. The anteclinal vertebra (the centre of motion) has a horizontal articulating facet anteriorly and an almost vertical one posteriorly and the posterior thoracic and lumbar vertebrae all have vertical articulating facets. Movement between these vertebrae is limited to dorsoventral flexion. The three epiaxial muscle systems combine in the lumbar region to form a very complex mass of muscles, the sacrospinalis. Dorsoventral flexion of the lumbar region is very well developed in swiftly moving carnivores, where it increases the length of the stride, for instance in greyhounds and cheetahs. It is also well developed in rodents and in many other small mammals. Large herbivores such as cattle, horses, and elephants usually have a rigid lumbar region.

Ventilation of the lungs in mammals involves movements of the ribs but the thoracic cavity is separated from the abdominal cavity by a muscular diaphragm whose movements are also of great importance. The hypaxial muscles of mammals are basically similar to those of the lizard.

In the Monotremata there are nineteen thoracico-lumbar vertebrae, of which sixteen bear ribs in *Tachyglossus* and seventeen in *Ornithorhynchus*. The transverse processes are short and the ribs articulate only with the centra.

In the Marsupialia the number of the thoracico-lumbar vertebrae is always nineteen, of which thirteen as a rule bear ribs.

In the Insectivora the number of the thoracico-lumbar vertebrae varies from nineteen in *Tupaia*, where there are thirteen thoracic and six lumbar, to twenty-four in *Centetes*, where there are nineteen thoracic and five lumbar. Small oval ossicles occur at the junction of the lower sides of the centra in the lumbar region. These ossicles are small in *Erinaceus*, larger in *Talpa*, and present in a few other Insectivora but in no other living mammals. They represent the remains of the reptilian intercentra (see p. 229).

In the Xenarthra the thoracico-lumbar vertebrae have an out-growth from the transverse process which articulates with the spine of the following vertebra. These additional zygapophyses are present in all forms, but reach their highest development in the Myrmecophagidae (anteaters), where second, third, and fourth pairs of extra zygapophyses are developed as the vertebrae are followed back. In the Dasypodidae (armadillos) the metapophyses are long and project upwards, helping to support the carapace. In the extinct *Glyptodon* almost all the vertebrae are fused together.

In the Cetacea the number of the thoracic vertebrae varies from nine in *Hyperoödon* to sixteen in *Balaenoptera*, and the lumbar from three in *Inia* to twenty-four or more in *Delphinus*. The transverse processes are well developed and they project as flat plates from the middle of the centra. The articulation of the trunk vertebrae is poorly developed in the Cetacea, in *Delphinus* for example the zygapophyses are practically absent in the lumbar region.

In both the Perissodactyla and the Artiodactyla the thoracico-lumbar vertebrae have very high neural spines and the centra are slightly opisthocoelous. In the Proboscidea there are twenty-three thoracico-lumbar vertebrae, of which nineteen or twenty bear ribs. A small number of lumbar vertebrae is found also in the Sirenia, which, as has already been pointed out, may be related to the Proboscidea. In *Manatus* there are seventeen thoracic and three lumbar, while in *Halicore* there are seventeen thoracic and two lumbar.

Sacral region

The sacral vertebrae are, by definition, those vertebrae which are united to the ilia. The union is in the first place by the rudiments of ribs and transverse processes, but complete fusion often occurs. There are usually only two or three vertebrae which fulfil this definition. The position of the sacral vertebrae is not fixed, even in one species, for the attachment of the pelvis shifts forwards during growth. In man, where this shift of the pelvis is well seen, the pelvis is in the embryo attached to vertebra thirty, but as development proceeds it becomes attached progressively to the more anterior vertebrae and detached from the posterior ones, so that, in the adult, the pelvis is usually attached to vertebra twenty-five or twenty-six. The actual position of the pelvis and so of the sacral vertebrae is therefore determined by conditions of growth. In *Cynocephalus* the pelvis grows backwards and the sacrum develops in a caudal direction, thus differing from all other mammals. The sacrum is formed by the sacral vertebrae fusing together. In addition to the sacral vertebrae proper, others from the caudal series may join in the fusion and so become pseudosacral vertebrae. Further strengthening of the sacral region is often effected by the complete fusion of the ilium with the sacral vertebrae, and often the ischium as well is fused with the pseudosacral vertebrae. This strengthening of the sacral region by fusion is an adaptive feature and it occurs in all groups of animals where a firm support is needed for the posterior part of the body. Almost all the variations in the sacral region may be seen in the Insectivora, an order in which the habits of the members vary widely.

In the Monotremata, *Ornithorhynchus* has two sacral vertebrae ankylosed together, while in *Echidna* there are three or four. In the Marsupialia there may be only one sacral vertebra (e.g. *Thylacinus*). In the kangaroos there are two sacral vertebrae which are ankylosed together. In the wombats four or five vertebrae unite to form the sacrum. In *Notoryctes* the fusion in the sacral region is more extensive, there being six vertebrae fused together and united not only with the ilia but with the ischia as well.

In the Insectivora, as has already been pointed out, the sacral conditions vary widely. In *Talpa* the vertebrae are firmly fused, in *Myogale* the vertebrae are fused and there is in addition a high crest formed of the fused neural spines, while in *Echinosorex* both the ilium and the ischium are united to the sacrum.

In *Cynocephalus*, where, as has already been mentioned, a backward shift of the pelvis occurs, the pelvis is, in the adult, only partly attached to the first sacral vertebra, so that, of the five so-called sacrals, the posterior two are caudals.

The Xenarthra show particularly extensive fusion of the vertebrae in the sacral region, the ischium as well as the ilium being united to the sacrum. The fusion is most marked in the armadillos.

Both the Artiodactyla and the Perissodactyla have only one sacral vertebra, to which the ilium is attached and a varying number of pseudosacral vertebrae are fused to it. In the Proboscidea there are two sacral vertebrae but they are not fused together.

In the Primates there are usually two or three sacral vertebrae with a certain number behind from the caudal series fused to them.

In the Sirenia the vestigial pelvis is attached in *Halicore* by a ligament to the transverse process of a vertebra which is not modified in any way; in *Manatus* the pelvis is not attached but lies below an unmodified vertebra, which is considered to be the sacral. The same condition as in *Manatus* occurs in the Cetacea. As both in the Cetacea and the Sirenia chevron bones are attached underneath the centra of the caudal vertebrae so distinguishing them from the thoracicolumbar, the position of the sacral vertebra may be fixed by its being immediately in front of the caudal series. There is no sacrum in either of these groups. The cetaceans swim by dorso-ventral movements of the tail which is expanded into lateral flukes supported by fibrous tissue. The three series of longitudinal muscles of the thorax continue into the tail and they (particularly the longissimus dorsi) are the elevators of the tail flukes. The rectus abdominis muscles continue as the hypaxial muscles of the tail and are the depressors of the tail flukes. Note that the movement of the tail of cetaceans is in the plane at right angles to that of the tails of Fishes, Amphibians, and Reptiles. Specialization of the lumbar vertebrae of the terrestrial ancestors of the whales meant the loss of horizontal flexibility at the posterior end of the trunk and the whales make use instead of the dorso-ventral flexibility characteristic of mammals. Whales have a very powerful diaphragm set obliquely. The Sirenia swim using fore-limb movements and their vertebral column is rather rigid and their body muscles are similar to those of terrestrial quadrupedal mammals.

Caudal region

The caudal vertebrae vary considerably in number and shape. When the tail is very long the anterior vertebrae possess all the characters of a normal vertebra, with well-developed neural spines, transverse processes, and zygapophyses, but towards the end of the tail the vertebrae are reduced to cylinders which represent the centra only. When the tail is short, degeneracy commences near the sacral region. As has already been noticed, the anterior vertebrae of the tail are often fused with the sacral to form pseudosacral vertebrae. Chevron bones, representing the haemal arches, are present in most of the orders of the Mammalia, though not necessarily in all the genera of any order. They are absent entirely in the Perissodactyla and the Artiodactyla.

In Primates the tail is very long in many forms. In the New World monkeys it is prehensile, but in the Old World forms it is never prehensile, even if it is long. From the fact that when monkeys are kept in captivity in cold climates the tails of the Old World monkeys are liable to get frost-bitten in winter, whereas those of the New World forms do not, we may assume that the tails of the Old World monkeys represent an organ which lacks full sensation and is degenerating. In the Anthropoidea the tail is reduced to four or five caudal vertebrae, which, in man, are fused together to form the os coccyx. In the Chiroptera the tail may be absent altogether, the vertebrae having completely fused with the sacrum.

16 *Pectoral and pelvic girdles, sternum, and limbs*

Fish

Pectoral girdle and pectoral fin

The pectoral girdle of fish has several functions and may originally have consisted of two parts: (1) a structure lying in the ventral musculature of the body and serving as a firm attachment for the pectoral fin and an origin for the muscles moving the fin; this is the scapulocoracoid, formed from cartilage bone; (2) a structure forming a firm posterior border to the pharyngeal region (which is perforated by the gill slits) and acting as the origin for the ventral (hypaxial) body muscles which extend posteriorly down the body and to the tail and also for the hypobranchial muscles which extend forwards and insert on the branchial arches ventrally. The former are used in swimming and the latter in feeding and respiration. In bony fishes, this structure consists of membrane bones and articulates dorsally with the skull.

Pelvic girdle and pelvic fin

The pelvic girdle always consists of cartilage or cartilage bone only. In fishes it is generally a simple transverse bar lying in the hypaxial musculature and serving as a firm attachment for the pelvic fin.

Dogfish and skate (Elasmobranch fishes)

No membrane bones are formed and the pectoral girdle consists only of scapulocoracoid but this includes among its functions those performed in bony fish by the membrane bones. The two girdles are fused midventrally (in the coracoid region) and the dorsal (scapular) portions lie close to the vertebral column embedded in the dorsal (epiaxial) musculature. In the skate, small supra-scapular cartilages unite the scapulae with the vertebral column on each side so that there is a complete ring of cartilage. This difference is presumably correlated with the flattened shape of the fish which swims using the expanded pectoral fins.

In the skate or dogfish the skeleton of the pectoral fin consists of three large proximal cartilages, the propterygium, mesopterygium, and metapterygium. The propterygium is the most anterior and smallest of the three. These three cartilages articulate with the glenoid cavity of the pectoral girdle, and bear distally a number of cartilaginous rods, the radialia. The radialia articulate distally with a number of polygonal cartilaginous plates, and beyond this the fin is supported by a number of dermal fin fibres, the ceratotrichia.

In the skate the propterygia and metapterygia are much elongated anteriorly and posteriorly in the horizontal plane. They are both jointed and the propterygium is united anteriorly to the olfactory capsule by the antorbital cartilage. The radials are extremely numerous, forming long jointed rods of cartilage lying parallel with one another and supporting the very large pectoral fins. Beyond the radials lie the dermal horny fin rays. Between the scapula and the propterygium lies a large scapulo-coracoid fontanelle through which passes the large brachial nerve.

In the dogfish and skate the pelvic girdle consists of a straight transverse bar of cartilage which sends off on each side a forwardly directed pre-pubic process from its lateral anterior edge. On its outer posterior border the pelvic girdle bears a pair of facets for the articulation of the pelvic fins.

In the pelvic fin there is only one large basal cartilage, the basipterygium. In both dogfish and skate the basipterygium projects directly backwards from its articulating facet with the pelvic girdle. Along its outer border lie a number of cartilaginous rods, the radials. The anterior radial in the dogfish and the first three in the skate articulate directly with the pelvic girdle. In both cases the anterior radial is larger than the rest. In the male the cartilaginous supporting bars of the claspers are attached to the posterior ends of the basipterygium.

Bony fish
(Fig. 39)

The cartilaginous girdle ossifies from two centres forming the scapula (dorsal) and the coracoid (ventral) with a broad glenoid cavity lying between them. The scapulocoracoid articulates with the posteromedian border of the cleithrum, a large membrane bone that forms the greater part of the membranous girdle. The ventral ends of the two cleithra meet below the gill region but do not fuse. Primitive bony fish, such as *Polypterus*, have a small clavicle on either side, ventral and anterior to the cleithrum and meeting in the mid-line. Above the cleithrum lies the supra-cleithrum, a short rod-shaped bone attached to the cleithrum below and to the post-temporal above. The latter may be considered either as part of the skull or as part of the girdle; it is forked anteriorly and articulates in Teleosts with the pterotic and epiotic bones (see p. 169). Loosely attached to the posterior border of the cleithrum above the scapula lies a long slender post-cleithrum which projects backwards, lying in the hypaxial musculature.

In Actinopterygian fish, the basal elements of the fins tend to be reduced or absent. In *Amia* the basalia of the pectoral fin are fused together to form a single plate and there are about ten small radials while in the cod there are no basals and the radials are reduced to four small 'brachial ossicles' which articulate directly with the glenoid cavity. The fin web is supported by bony fin rays called lepidotrichia. These are forked proximally and articulate with the radials. The muscles which move the fin insert on to the proximal ends of the rays.

The pelvic girdle is small and generally consists of a triangular plate united anteriorly by a ligament to that of the other side. There are no basals and the radials are either reduced to small knobs, as in the salmon, or absent, as in the cod, where the fin-rays articulate directly with the girdle. In more advanced Teleosts, such as the cod, the pelvic girdle has moved forwards and articulates with the cleithrum, anteriorly and ventrally to the scapulocoracoid.

Tetrapods

The majority of terrestrial vertebrates walk using four limbs in a diagonal pattern. The hind limbs propel the body forwards during the process of retraction. The fore limbs act as supports for the anterior end of the body and act as brakes. The hind limb and pelvic girdle are very well developed and there evolved early a connection between the pelvis and the sacral region of the vertebral column through which the thrust of the limb is transmitted to the bulk of the body.

The gill-slits are closed in all air-breathing Tetrapods and the pectoral girdle has lost its connection with the back of the skull. A neck may be present. The pectoral girdle is never directly connected with the vertebral column and can always be moved relative to'it. It is much more variable in structure than the pelvic girdle. Compared with fishes, there is a reduction in the cleithrum and an increase in area of the scapulocoracoid. Two new median ventral bones are developed—the interclavicle and the sternum.

The salamander

Pectoral girdle
(Figs. 16, 17, 18)

All the membrane bones are lost and the girdle remains cartilaginous throughout life. It differs from the cartilaginous girdle of the fish in possessing a precoracoid process that projects forwards from the coracoid in the region of the glenoid cavity. The two coracoids form large flat plates which overlap ventrally in the middle line. The scapula is a narrow elongated plate of cartilage projecting upwards above the transverse process of the vertebrae. The girdle may be slightly ossified around the glenoid cavity.

A small median cartilaginous plate, the sternum, lies in the middle line and overlaps the posterior part of the coracoids.

The two girdles overlap ventrally but each can be moved independently of the other relative to the body muscles which lie internal to them. Muscles causing such movements include the trapezius, rhomboideus, and serratus muscles. The sternum is the origin for the sternohyoideus muscle (a hypobranchial muscle), passing forward to insert on the hyoid plate, and for the rectus abdominis muscle (part of the hypaxial musculature) passing posteriorly to the pelvic girdle.

In the Anura the cartilaginous girdle is more completely ossified into coracoid and scapula bones. Of the membrane bones deposited round the cartilaginous girdle in the bony fish, only one remains, the clavicle, which wraps round the precoracoid. The sternum is short and partly ossified.

In frogs, the clavicles and coracoids are fused with the sternum so that the two girdles form a single functional unit and are not pulled apart when the shock of landing is taken by the fore limbs. In toads, the girdles are not fused to the sternum but the two coracoids are tied together anteriorly by a ligament. Each is free to rotate about this connection but the two cannot easily be pulled apart. The mechanics of this girdle resemble that of the lizard (see p. 245) but the movement is less accurately controlled. The toad walks on land whereas the frog leaps.

Pelvic girdle   The pelvic girdle consists of a flat plate lying in the ventral body wall and connected to the sacral vertebra by a dorsally directed process. The pelvic girdle is incompletely ossified. The dorsal portions attached to the sacral vertebra form the paired ilia and the posterior ventral parts form the paired ischia. The anterior ventral part of the girdle is unossified but forms the paired pubic bones in other forms. The pubic cartilage is pierced on either side by the small obturator foramen and, in front, in the middle line there projects forwards a bifid cartilaginous plate, the epipubis. The cup-shaped acetabulum serves to receive the head of the femur.

In the Anura the three paired bones of the pelvic girdle are more completely ossified and the epipubis is absent. This girdle is very specialized with the long ilia directed forwards to the sacral vertebra, the small ischia and pubes and the close approximation of the two acetabula. The hind legs are large and are simultaneously retracted when frogs leap and when toads swim.

Limbs
(Figs. 16, 17, 18, 19)

The paired limbs of the Amphibia are pentadactyl limbs which are common to all the land vertebrates. Each limb is composed of a single proximal bone, the humerus or femur, which is articulated to a pair of bones, the radius and ulna or tibia and fibula. To these bones are attached a row of not more than three carpal or tarsal bones and these articulate with a distal row of not more than five carpals or tarsals which are separated from the proximal row by a small bone, the centrale. On to the distal row of carpals or tarsals are articulated not more than five digits, which are composed of one proximal metacarpal or metatarsal and from one to four phalanges.

It is customary to number the digits from the inside outwards. The first digit is termed the pollex or thumb on the fore limb and the hallux on the hind limb.

With the exception that in the salamander the hallux and pollex are missing and in the newt, pollex only, the hind and fore limbs of the newt and salamander approximate very closely to the primitive type described above. The humerus has at its proximal end a round head which articulates with the glenoid cavity of the pectoral girdle. Along the anterior border there runs a strong ridge, the deltoid ridge, which serves for the attachment of the arm muscles. The radius and ulna are both small bones.

The femur has at its proximal end a rounded head which articulates with the acetabulum of the pelvic girdle. The ridges for the attachment of the muscles are not strongly marked.

The musculature of the frog is described on p. 70. The hind limbs of Anura are very specialized.

The lizard
(Fig. 49)

The pectoral girdle of the lizard is much more ossified than that of the Amphibia. It is much larger and is closely associated with the large sternum. The cartilaginous portion of the girdle ossifies to form the scapula above and the coracoid below, the two forming a single functional unit. The scapula is divisible into two portions, the upper of which, the suprascapula, is cartilaginous, and the lower, the scapula proper, is ossified. The ventral end of the scapula forms a portion of the glenoid cavity. The coracoid is ossified and fenestrated. It is sutured to the scapula and contributes to the glenoid cavity. The postero-ventral border of the coracoid is curved and fits into a groove in the sternum, the coracosternal groove. The sternum is a broad plate lying in the mid-line with the two coracoids fitted into its anterior edges, the coracosternal grooves. It is hardened by calcification but true ossification does not as a rule take place. To the lateral edges of the sternum are united the ends of the five anterior ribs.

Strongly attached to the mid-line of the sternum is the T-shaped interclavicle (a membrane bone) with its anterior and lateral arms in front of the sternum. The clavicles meet in the mid-line and are attached to the anterior end of the interclavicle. The clavicles are slender bones lying along the front edges of the coracoids and they articulate laterally with the suprascapulae. In some lizards, the lateral arms of the interclavicle also articulate with the suprascapulae but in others (see Fig. 49) these arms are short and bound to the clavicles.

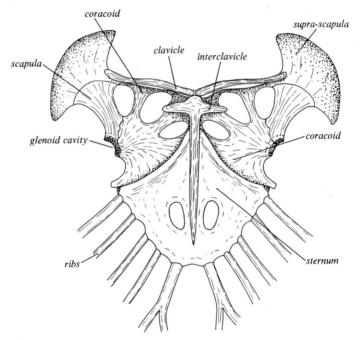

FIG. 49. Ventral view of the pectoral girdle of the lizard.

In dried specimens there appears to be little possibility of movement in this girdle, but in a fresh lizard the two scapulo-coracoids can pivot independently and there are muscles that produce these movements. The fulcrum for the rotation of the scapulocoracoid is the articulation between the clavicle (sometimes also the interclavicle) and the suprascapula; the plane of the movement is controlled because the medial edge of the coracoid must slide in the coracosternal groove. The effect of the rotation is that the glenoid cavity is moved forwards and backwards in relation to the rest of the body and this leads to a slight lengthening of the stride as the fore limb is swung forwards and then pulled backwards. The clavicles and interclavicle brace the glenoids apart.

This movement within the pectoral girdle is probably an ancient feature of the Tetrapods. It is found also in the toad (see p. 243) although the sternum is poorly developed and there is no interclavicle. Reptiles that have this movement usually have a T-shaped interclavicle and well-developed clavicles; typically, their fore limbs are held out laterally at the sides of the body. In Reptiles whose limbs are held beneath the body, as in chameleons and crocodiles, the coracosternal groove is replaced by a suture that allows bending but not sliding and the interclavicle is reduced or lost; the clavicles may also be lost.

Pelvic girdle

The pelvic girdle is completely ossified. It consists of three pairs of bones which on either side form the os innominatum. The ilia, the ischia, and the pubes occupy the same relative position as in the salamander, but the large pubo-ischiadic foramina separate the pubic from the ischial region. The three bones meet in the acetabulum, a concave pit that lodges the head of the femur. The ilium is an irregular rod-shaped bone that projects obliquely backwards from the acetabulum. At its upper end it is attached by a strong ligamentous connection to the transverse processes of the sacral vertebrae. The pubis projects forwards and downwards from the acetabulum. It meets with the pubis on the other side to which it is joined by the pubic symphysis. From the anterior border of the pubic symphysis there extends forwards the cartilaginous epipubis. Laterally each pubis has a conspicuous process projecting forwards, the processus pubis.

The ischium projects inwards and downwards behind the pubis to meet its fellow in the mid-ventral line. The ischia are united in the middle line by a symphysis in the same way as the pubes. Stretching between the pubic and ischiatic symphyses is a band of ligament. The spaces on each side of this ligament, bounded by the pubes in front and the ischia behind, are the pubo-ischiadic foramina (= thyroid fenestrae). Each pubis has a small obturator foramen for the passage of the obturator nerve to the pelvic and thigh muscles. In life, the puboischiadic foramen is filled with strong connective tissue acting as the origin for a muscle.

The principal retractor muscle of the femur—the principal propulsive muscle of the lizard—is the caudofemoralis muscle. This arises from beneath the caudal vertebrae and is inserted on the femur on to a ridge or knob situated ventrally and called the fourth trochanter. The postero-dorsal inclination of the ilium may reflect stresses imposed on the iliosacral junction when the caudofemoralis muscle contracts. Strong ventral muscles arise from the pubis and ischium and are inserted near the head of the femur into a fossa and on to the internal trochanter. These enable the lizard to lift its body from the ground.

Limbs

The limbs of the lizard do not depart widely from the general pentadactyl type. The ends of the digits are furnished with claws. The phalangeal formula is 2.3.4.5.3. When the lizard is at rest, the limbs are splayed out laterally with the under surface of the feet applied to the ground.

The turtle

The evolution of the box-like armour of the Chelonia has involved considerable changes in the skeleton of the trunk region. The armour consists of bony plates that are covered by skin with horny plates (the tortoise-shell) whose pattern is not identical with that of the bones below. The bones form a dorsal shield, the carapace and a ventral shield, the plastron, usually fused together laterally and widely separated anteriorly, for the neck and fore limbs, and posteriorly, for the hind limbs and tail.

The carapace typically consists of eight median neural plates, fused with the neural spines of eight trunk vertebrae which are also immovably fused with each other. Lateral to these are eight costal (or pleural) plates which are fused to the ribs which also are immovably fused with the vertebrae. Lateral to the costals are the marginal (or peripheral) plates. There is a median nuchal plate anterior to the first neural and one or more median pygal plates posterior to the last neural.

The plastron consists of a median plate, the entoplastron, probably equivalent to the interclavicle, with a pair of plates, the epiplastra, anterior and lateral to it and three pairs of plates posterior to it, the hyo-, hypo- and xiphi-plastra. The epiplastra probably represent clavicles and the other six ventral plates are probably derived from abdominal ribs (gastralia). The plastron is usually firmly sutured laterally to the marginal plates of the carapace. Joints may be present in the plastron, as in box turtles. The armour is reduced in some Chelonia, e.g. *Trionyx*. The leathery turtle has aberrant armour.

The position of the pectoral and pelvic girdles of Chelonia

is unique for they lie internal to the ribs (here part of the armour). During the development of the embryo, the shell extends forwards and backwards and thus encloses the girdles.

Pectoral girdle

The clavicles and interclavicle are probably part of the plastron and the sternum appears to be absent so that the movable girdle consists only of the original cartilage elements, the scapula and coracoid. The dorsal blade of the scapula articulates with the first costal plate and there is a well-developed acromion process which articulates with the epiplastron (= clavicle). The coracoid forms part of the glenoid and then is directed ventro-medially, posterior to the acromion process. The whole girdle is thus tri-radiate and fits between the carapace and plastron.

Pelvic girdle

This is similar to that of the lizard in structure, with the pubes and ischia each meeting its fellow ventrally but separated by the pubo-ischiadic foramen (= thyroid fenestra) on each side. The ilia articulate with two sacral vertebrae (whose neural spines are not fused with the carapace) and also with the posterior costal plates. In the Pleurodira, the pelvic girdle is fused to the carapace and plastron.

Limbs

These are basically similar to those of the lizard but the phalangeal formula is 2.3.3.3.3.

The crocodile

Pectoral girdle

The scapula and coracoid are rather elongate and meet at an angle at the glenoid fossa. The coracoid meets the sternum along a suture line which allows bending but there is no sliding joint as in the lizard. There is a splint-like interclavicle attached to the front of the sternum in the mid-ventral line. The interclavicle lacks lateral arms and there are no clavicles. This girdle is more like that of the Bird than that of the lizard but differs from the Bird in the absence of clavicles and the presence of an interclavicle.

Pelvic girdle

The acetabulum of crocodiles is perforate and is formed by the ilium and ischium only, the pubis being excluded by the ischium. The two pubes extend forwards as cartilaginous epipubes which are connected to each other by a strong ligamentous membrane. This membrane extends forwards to the last pair of abdominal ribs and backwards as a mid-ventral ligament to the cartilaginous ventral symphysis of the ischia. The ilium has a broad flange and its articulation to the first sacral vertebra is directly above the acetabulum.

Limbs

The crocodile rests with its limbs sprawled out sideways but it runs with them held beneath the body and with the elbow rotated backwards and the knee rotated forwards as in its probably bipedal ancestors. The front legs are shorter than the hind legs. The musculature of the limbs and girdles is basically similar to that of the lizard but differs in detail.

There are five digits in the fore limb but the three inner ones are stronger and bear claws. The carpals are reduced to two elongated proximal elements and a short distal element and a pisiform is present. There are four digits in the hind limb, the three inner ones bearing claws. The fifth digit is represented by a small hooked metatarsal. The astragalus is closely applied to the tibia and there is an intratarsal joint which foreshadows that in birds.

The Bird *(Gallus)*

Pectoral girdle
(Figs. 26(a), (b),
27)

The pectoral girdle of the bird is very greatly modified in consequence of the adaptation of the fore limbs for the purpose of flight. The scapula is long and slender, and projects backwards parallel to the vertebral column over the dorsal side of the ribs as far as the pelvic girdle. The coracoid is placed at an angle of about 90° to the scapula. It is a strong cylindrical bone, broad at its ventral end where it is attached to the sternum. The glenoid cavity is formed as usual at the union of the coracoid and scapula on the outer side of the girdle. Projecting downwards from the dorsal end of the coracoids is a V-shaped bone, the furcula. This is formed by the clavicles united in the middle line ventrally. The ends of the furcula, coracoid, and scapula meet beside the glenoid cavity and enclose a large foramen, the foramen triosseum, through which passes the tendon of the supracoracoideus muscle (see Fig. 26(b)).

Sternum and ribs
(Figs. 26(a), (b),
27)

The sternum is a large structure extending backwards to cover nearly the whole of the abdominal region. In the middle line the sternum bears a strong keel which serves for the attachment of the pectoralis muscles; anteriorly and posteriorly it is produced into lateral processes, the costal and xiphoid processes. Between these processes are attached the ventral ends of the ribs.

Pelvic girdle
(Fig. 50)

The pelvic girdle is a large structure which is firmly ankylosed with the vertebral column. It is difficult to distinguish the sutures of the three bones which form the os innominatum. The ilium is the largest bone of the pelvis and forms a plate extending for a considerable distance alongside the vertebral column both in front and behind the acetabulum. The ilium is continuous backwards with the ischium and is indistinguishably fused with it. The ischium is also ankylosed to the sacral vertebrae. Behind the acetabulum is a large foramen in the ischium, the iliosciatic foramen. The pubis is a small slender strip of bone that extends from the ventral border of the acetabulum along the lower edge of the ischium and projects backwards along the sides of the body cavity. In front of the acetabulum is a small process, the pectineal process, which may possibly correspond to the processus pubis of Reptiles. There is no pubic or ischiadic symphysis in the middle line.

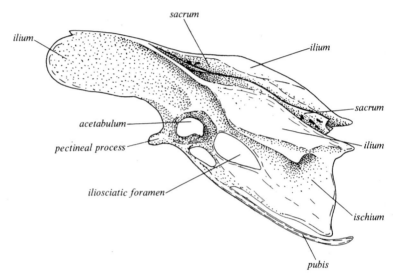

FIG. 50. Lateral view of the sacrum and pelvic girdle of a fowl.

Fore limb
(Fig. 51)

The fore limb of the bird deviates considerably from the typical pentadactyle limb with the formation of the wing for flight. The main peculiarities of the skeleton are (1) the complete absence of the two outer digits and (2) the fusion of the distal row of carpals with the metacarpals to form a compound bone, the carpometacarpus. When the limb is extended its position resembles that of primitive vertebrates, the preaxial border being directed forwards. When the wing is folded the upper arm and hand are bent upon each other with the elbow directed backwards and the wrist forwards. All the bones of the limb are strong but light, and contain air spaces.

The humerus is elongated and expanded at both ends. The proximal convex head articulates with the glenoid cavity and distal to it are two ridges for muscular attachment on the pre- and postaxial borders of the humerus. The preaxial ridge is known as the deltoid ridge, or greater tuberosity, and on it are inserted the pectoralis and supracoracoideus. The post-axial ridge, or lesser tuberosity, is the larger of the two and has at its posterior surface a deep pit, the pneumatic foramen, leading to an air cavity in the shaft of the bone.

The radius and ulna of the forearm are well developed and separate from each other. The radius bears a proximal cup-shaped articular surface for the preaxial condyle of the humerus. The ulna is longer and stouter than the radius, and bears the secondary feathers of the wing. It bears proximally the large articular surface for the postaxial tubercle of the humerus, beyond which it projects as the olecranon process. Distally it articulates with the carpus and the radius.

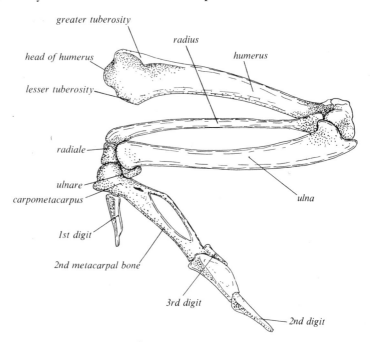

FIG. 51. Left fore limb of a fowl.

The carpal bones are represented by a proximal row of two bones, the radiale, articulating with the radius, and the ulnare, the larger of the two, articulating with the ulna. In the embryo the distal row of metacarpals is formed by three separate bones corresponding to the inner three carpals of the series. In the adult they fuse together with the three metacarpals of the inner three digits to form an elongated bone, the carpometacarpus, in which can be distinguished the three united metacarpals. The hand is very long, almost as long as the forearm. The second and third digits bear the long primary feathers. The wrist joint lies between the proximal row of carpals and the carpometacarpus bone. There is no general flexion possible between both rows of carpals and at the articulation of the carpals with the forearm and hand as in other vertebrates.

The first digit or pollex has two phalanges, the terminal one being very small. It bears the bastard wing. The second digit is the largest of the three and has three phalanges, the proximal one being large and flat and the distal one small. The third digit has only a single phalanx which is closely united with the proximal phalanx of the second digit at either end.

Hind limb
(Fig. 52)

The hind limb of the bird is greatly enlarged for bipedal locomotion, and the bones are extremely strong although perforated by air cavities. All the bones are elongated and the fusion of the tarsal bones with adjacent portions of the limb has been carried farther. The tarsometatarsus is formed in the same way as the carpometacarpus, but is greatly elongated and raised off the ground so that only the digits touch the ground. The femur is elongated, but is often the shortest bone in the leg. Proximally it bears the inwardly directed head articulating with the acetabulum. On its outer surface it bears a process, the great trochanter, for the attachment of muscles. Between the head and the great trochanter the end of the femur articulates with the ilium. A very strong joint is thus formed, capable of supporting the animal, even though the centre of gravity lies well in front of the acetabulum. The distal end of the femur slopes downwards and forwards, and bears a deep groove for the tendon of the extensor muscle of the leg. The patella bone lies in this tendon. This groove is bounded laterally by two ridges or condyles that articulate with the tibia, the outer condyle also articulating with the fibula. The whole of the femur is embedded in the body musculature and only the knee projects from the surface.

The tibia is very long and straight, considerably longer than the femur, and is fused with the proximal row of tarsal bones forming the tibiotarsus. Proximally it bears two facets for articulation with the femur, and round its anterior face it bears the prominent cnemial crest, on to which is inserted the extensor muscle of the thigh beyond the patella. Laterally another crest is present for muscular attachment. On the distal end of the bone is the pulley-like articular surface for the tarsometatarsus.

The fibula is a very slender bone closely applied to the outer

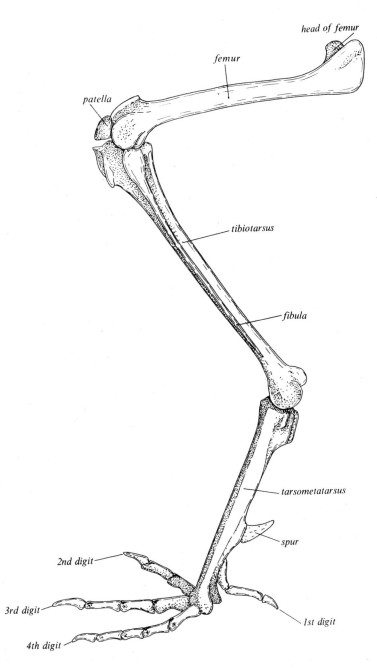

FIG. 52. Left hind limb of a fowl.

surface of the tibiotarsus, with which it is often partially fused. Proximally, it enlarges and articulates with the femur, and the slender distal end usually does not reach as far as the ankle.

The single ankle joint lies between the tibiotarsus and the tarsometatarsus and is directed backwards.

The tarsometatarsus is a long stout bone about the length of the femur. Its proximal end bears two cup-shaped facets for the tibiotarsus, and posteriorly it bears a prominent ridge traversed at its base by the longitudinal canal for the passage of the tendons to the flexor muscles of the toes. Behind this ridge two sesamoid bones[1] are usually present. In the shaft of the bone can be distinguished three fused metatarsals which separate from one another distally, each ending in a pulley-like facet for articulation with the corresponding digit. In the male bird there is on the inner side of the tarsometatarsus a bony outgrowth ending in a sharp point. This is the 'spur', a secondary sexual character, which is only fully developed in old birds.

The first four digits are present, the fifth being absent. The hallux is connected with the first metatarsal and is directed backwards. It has two phalanges, the last one bearing a claw. The second, third, and fourth digits are directed forwards, slightly diverging from one another. They are composed of three, four, and five phalanges respectively, the distal phalanx bearing the claw.

The Mammalia   Primitive Reptiles stood with their limbs held out laterally and with the femur and humerus parallel with the ground; they walked using lateral undulation of the body as well as protraction and retraction of the limbs. Mammals, however, stand with the limbs beneath the body and they walk without lateral undulation; many show dorsoventral flexion in the lumbar region during locomotion. The change in attitude of the limbs is accompanied by considerable changes in anatomy, especially in the pectoral girdle. The fore limb has rotated so that the elbow joint is directed backwards instead of outwards and the hind limb is rotated so that the knee joint is directed forwards. In the fore limb, a further rotation of the bones of the fore-arm round each other in the opposite direction to the elbow rotation has brought the pollex to the anterior inner side of the foot. In the hind limb, the original preaxial border lies on the inner side after the knee rotation and the hallux is directed forwards and inwards. The phalangeal formula of Mammals is basically 2.3.3.3.3 and both sets of digits are anterior to the carpal/tarsal joints.

[1] Sesamoid bones are small nodules of bone developed in the tendons of some muscles close to their insertion.

In the MONOTREMES the pectoral girdle is like that of the lizard. It is, however, almost completely ossified and the sternum is jointed. There is a large T-shaped interclavicle attached at its posterior end to the presternum. The clavicles are bound to the anterior borders of the arms of the interclavicle and each articulates laterally with the acromion process of the scapula. The scapula is a flat bone with raised edges for the attachment of muscles but there is no spine (as in other mammals). There are two ossifications in the coracoid region, called precoracoid and coracoid; this condition is characteristic of the synapsid Reptiles compared with the single coracoid ossification of diapsid Reptiles. The coracoid forms a portion of the glenoid cavity, the other portion being formed, as usual, by the scapula. The median edges of the coracoid and precoracoid insert into a groove in the presternum. Thus the scapulo-coracoid unit on each side can rotate about the articulation between scapula, clavicle, and interclavicle in the same way as in primitive Reptiles. Monotremes hold their limbs laterally from the body with the humerus horizontal as in lizards. The supracoracoideus muscle arises from the ventral surface of coracoid and precoracoid as in lizards.

In all other mammals, the interclavicle is lost. The precoracoid may sometimes be represented by small vestiges. The coracoid is reduced to form a 'process' of the scapula beside the glenoid cavity. The scapula has on its outer surface a prominent ridge, the spine, which projects at its lower end as the acromion process; there may be on this an additional process, the metacromion. The spine and the infraspinous fossa behind it represent the scapula of Reptiles and Monotremes while the supraspinous fossa in front of the spine is a new development. The supracoracoideus muscle of Reptiles has become the supraspinatus and infraspinatus muscles of mammals and they arise from the scapula.

The clavicle articulates with the acromion process of the scapula laterally and with the presternum medially. The clavicles are present as a rule only in arboreal forms and in animals which use the fore limb in a variety of ways. The clavicle is absent or is reduced in all forms where the anterior limbs perform movements which only entail their swinging to and fro more or less in the line of the long axis of the body, as in the Ungulates, and the Carnivora. The clavicle is also absent in the aquatic Cetacea and Sirenia.

The sternum in the Mammalia is a jointed structure and may be divided into three portions, the presternum at the anterior end, which is single, the mesosternum, which may be subdivided into 'sternebrae', and a posterior single portion, the xiphisternum. As a rule the first pair of thoracic ribs joins the presternum, the succeeding ribs are attached between the sternebrae, and the last pair of ribs to reach the sternum is attached to the xiphisternum.

In the INSECTIVORA there are considerable remains of the coracoid bones, mainly of the sternal ends. The clavicle is well developed, except in *Potamogale*, where it is absent. It is connected with the sternum and the acromion process by vestiges of the precoracoid. In the mole the scapula is high and narrow, without a prominent spine or acromion. The clavicle is of irregular shape and articulates with the presternum and humerus and is connected by ligaments with the scapula.

In the CHIROPTERA a slight keel is developed on the sternum and the clavicles are well developed.

In PRIMATES vestiges of the precoracoid occur at either end of the clavicle which is well developed.

In the CETACEA the scapula has a characteristic shape, being flat and fan-shaped. The spine is low and the acromion and coracoid processes are much flattened.

In the XENARTHRA, both in the anteaters and the sloths, the anterior border of the scapula is produced over the coracoscapular notch and fuses with the coracoid process. Both coracoid process and acromion process are large. In *Myrmecophaga jubata* (giant anteater) the coracoid is a separate bone. In sloths and armadillos the clavicle is well developed.

In the RODENTIA the shoulder girdle is rather primitive and there are remains of the precoracoid at either end of the clavicle. Remains of an epicoracoid are also to be seen lying against the presternum.

Pelvic girdle

As in Reptiles, the pelvic girdle of mammals consists of three bones on each side, the ilium, ischium, and pubis, together forming the os innominatum. The ilium is immovably attached to the sacral vertebrae and the pubis and ischium each meet their fellow ventrally. The obturator foramen on each side is enlarged so that the pubis and ischium are partly separated. This enlarged obturator foramen superficially resembles the pubo-ischiadic foramen of the lizard and turtle but those two still retain a separate foramen for the obturator nerve. All three bones take part in the formation of the cup-shaped acetabulum, which receives the head of the femur.

The principal retractor muscles of the femur are the gluteal muscles assisted by the flexor cruris group of muscles (see rat, p. 141) and these all arise from the ilium, sacrum or ischium. The caudofemoralis muscle of mammals is small. The acetabulum of mammals is situated well behind the union of the ilium with the sacral vertebrae in striking contrast to that of lizards and the difference is probably related to the difference in origin of the retractor muscles. The ilium of mammals is well placed to convey the force of the hind limb pushing the body forwards via the sacrum and vertebral column. In mammals with powerful gluteal muscles, such as horses, cattle, elephants, and the gorilla, the ilium is much expanded.

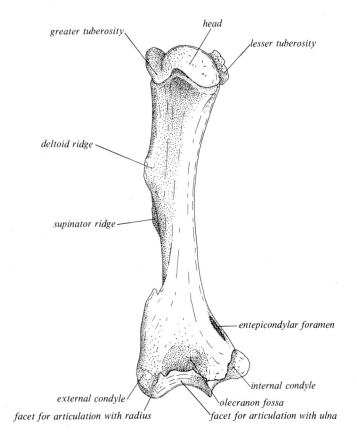

FIG. 53. Posterior view of the humerus of a kangaroo.

The chief difference from the usual type of pelvis occurs in Monotremes and Marsupials where a pair of slender epipubic bones is attached to the anterior border of the pubis in both sexes. These support the abdominal muscles and the pouch, if present. In *Tachyglossus* the acetabulum is perforated, a feature found in no other mammal and reminiscent of birds. In the Insectivora the symphysis is sometimes non-existent, and always weak, as in *Erinaceus* for example, where it is confined to the pubis. In the Edentates the symphysis is short, and in the sloths, armadillos, and anteaters the pelvis is united with the vertebral column both by the ischia and ilia. In Carnivora Fissipedia the pelvic symphysis is long, but in the Pinnipedia it is short or non-existent. In the Chiroptera the pelvis is firmly united with the sacrum, both by the ilia and the ischia, and there is, as a rule, no ventral symphysis. In the Sirenia and the Cetacea the pelvis is vestigial and is represented by two slender bones lying below the vertebral column at a point just in front of where the development of chevron bones commences. There is no trace of an acetabulum in the pelvis of these groups, and only in *Halicore* are the bones representing the pelvis attached to the vertebral column.

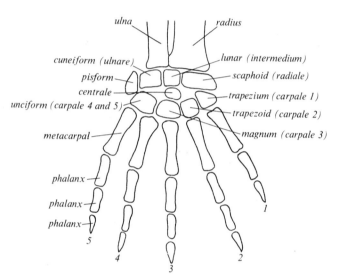

FIG. 54. Diagram of the manus showing the typical arrangement of the carpal bones in mammals.

The limbs
(Figs. 53–7)

Both the fore and hind limbs are constructed on the basic pentadactyl plan described on p. 244. The general relations of the bones of the hand and foot can be seen in Figs. 54 and 57. The distal phalanges bear either a nail or a claw.

THE FORE LIMB. The humerus articulates with the glenoid cavity by a large head. On the inner side and at the same end as the head is the lesser tuberosity, while on the outer side is the greater tuberosity. Between these two tuberosities is the bicipital groove. The greater tuberosity is continued down the humerus on the outer surface as a roughened ridge, the deltoid ridge. The tuberosities and the ridges vary considerably in their development in different orders. They are perhaps most clearly seen in the Marsupialia (see Fig. 53). At the distal end

of the humerus is the trochlear groove, where it articulates
with the radius and the ulna. On either side of the groove are
the internal and external condyles. Above the internal condyle
there is, in many mammals, the entepicondylar foramen. Passing upwards from the external condyle is the ectocondylar or
supinator ridge. In front and behind the trochlear are two
fossae, of which the posterior, the olecranon fossa, is by far
the deeper. The olecranon fossa receives the olecranon process
of the ulna when the arm is extended. In some forms these
two fossae communicate with each other by a foramen, the
supracondylar foramen.

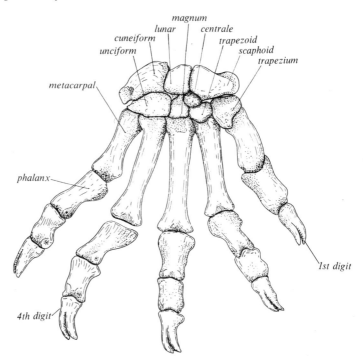

FIG. 55. Left manus of a Creodont, *Patriofelix ferox*, showing free scaphoid and
lunar and the presence of a free centrale.

Of the two bones forming the forearm the radius lies slightly
in front of the ulna. It is external to the ulna at its proximal
end, interior to the ulna at its distal end. The proximal end
of the ulna forms the hinge joint of the elbow. At this end
there is the deep sigmoid notch into which the trochlear of the
humerus articulates, while projecting beyond the notch is the
olecranon process of the ulna. The distal end of the ulna tapers
and, except in the elephant, is always smaller than the distal

end of the radius, which is expanded and forms the larger portion of the carpal joint (see Figs. 54 and 59).

In primitive forms the radius and the ulna are separate bones capable of a certain amount of rotation on each other. In Rodentia, for example, this power of rotation is lost and the radius and ulna are fixed in one position. In cursorial forms the radius is well developed and the ulna is reduced. The ulna may not extend the whole length of the radius but always remains at the proximal end forming the olecranon process.

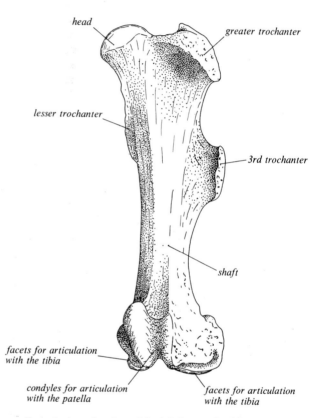

FIG. 56. Anterior view of the left femur of a rhinoceros.

Of the carpal bones, the primitive arrangement of which is shown in Fig. 54, the proximal row consists of three bones and the distal row of four, with a centrale bone in the middle between the two rows. On the outside, between the unciform and the cuneiform a sesamoid bone, the pisiform, may be developed in the tendon of the flexor muscles. The presence of a centrale is a primitive feature seen for example in Creodonts (Fig. 55) and in many Primates. Fusion of the bones often occurs.

THE HIND LIMB. The hind limb consists of the femur or thigh bone and two bones articulating with its distal end, the tibia and fibula. The ankle joint is formed of the tarsal bones while the metatarsals and phalanges form the skeleton of the foot and toes.

In the hind limb the femur corresponds in position with the humerus. The proximal end of the femur bears on its inner side a large rounded head which articulates with the acetabulum (see Fig. 56). On the outer side, at the proximal end, is a large roughened projection, the great trochanter. The great trochanter is separated from the head by a notch. On the inner side just below the head is another smaller roughened surface, the lesser trochanter. On the outer surface, usually about one-third of the femur length from the proximal end, is the third trochanter. The great trochanter is the principal insertion of the gluteal muscles and the lesser trochanter is that for the iliopsoas muscles (although in a similar position, it is not equivalent to the internal trochanter of Reptiles). The third trochanter is the insertion for part of the gluteal muscles and is present only in certain orders, e.g. insectivores, rodents, and perissodactyls. The distal end of the femur has two condyles which articulate with the tibia, it is also grooved on its anterior face for the reception of the tendon in which the patella lies.

Corresponding to the radius and ulna in the forearm are the tibia and fibula. The fibula is a slender bone and takes no part in forming the knee joint and very little, sometimes no, part in forming the ankle joint. The tibia is a large bone expanded at both ends. There are two facets at the proximal end for articulation with the femur and a notched surface at the distal end for articulation with the astragalus. The tibia and fibula are often ankylosed together. The anterior surface of the knee joint is covered by a sesamoid bone, the patella or knee-cap.

The arrangement of the bones in the tarsus (see Fig. 57) is much the same as that in the carpus. There are, however, only two bones in the proximal row, the two bones corresponding to the scaphoid and the lunar having fused into one bone, the astragalus. The bone corresponding to the centrale, the navicular, is always well developed. The ankle joint is always between the tibia and the tarsus and never between two rows of tarsal bones as in the birds and some reptiles. Sesamoid bones are often present and, of these, the pisiform is the most frequent.

The following are the chief modifications of the fore and

hind limbs from the primitive plantigrade arrangement just
described.

Monotremata

In the Monotremata the humerus is peculiar in shape owing
to the development at either end of flat expansions for
the attachment of muscles. An entepicondylar foramen is
present. The ulna has a very large and expanded olecranon
process. In the carpus there is no centrale, but there are several
other sesamoid bones in addition to the pisiform. The tarsus
is also equipped with several sesamoid bones. The fibula has
a process resembling the olecranon process of the ulna, a
unique feature. In the male there is a curious sesamoid bone,
which is articulated to the tibia and bears the horny spur.

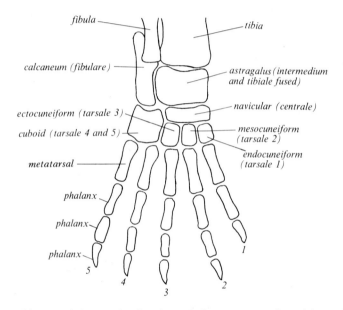

FIG. 57. Diagram of the pes showing the typical arrangement of tarsal bones in
Mammals.

Marsupialia
(Figs. 53, 58)

In the Marsupialia the humerus (see Fig. 53) has an entepicondylar foramen in all forms, except *Notoryctes*, and the
ridges are well marked. In *Notoryctes* the ulna has an exces-
sively long and curved olecranon process for the very powerful triceps muscle used in digging. In the wombat the humerus
is expanded in a manner recalling the condition found in the
Monotremata. The carpus has no centrale and the lunar is
usually small and sometimes absent. The femur has no third
trochanter. In the Diprotodontia, the foot exhibits syndactyly,

with the second and third digits very slender and enclosed in a single integument; they are used for combing the fur. In the wombat, the toes are less reduced but partly connected by skin. In the kangaroos, the hind limbs are much larger than the fore. The tibia and fibula are greatly elongated and so is the metatarsal of the fourth digit. The pollex is absent and digits two and three are syndactylized as in other Diprotodontia.

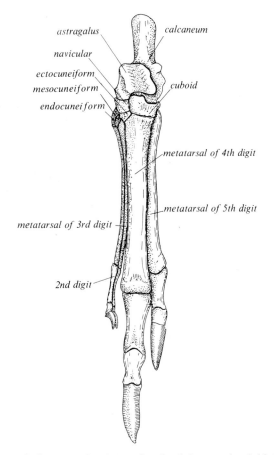

astragalus
navicular
ectocuneiform
mesocuneiform
endocuneiform
calcaneum
cuboid
metatarsal of 4th digit
metatarsal of 5th digit
metatarsal of 3rd digit
2nd digit

FIG. 58. Left pes of a kangaroo showing syndactyly of the second and third digits.

Insectivora

In the Insectivora the wide range of habitats occupied by the group necessarily entails a certain divergence from the general type in some forms, but the divergence is nowhere very considerable and one or more primitive characters remain. The humerus has an entepicondylar foramen (except in

Erinaceus) and a supracondylar foramen. The ridges and the tuberosities are better developed in burrowing forms than in the others. The mole has a very peculiar humerus, being very short and flattened, with curved expansions at either end. The radius and the ulna are separate and the olecranon process is very large. The hand is usually complete with five digits and there is often a centrale bone (as in *Talpa* but not in *Erinaceus*). The hand in the burrowing forms shows considerable modification in the direction of broadening and the development of strong ungual phalanges. In the hand of the mole there is a large radial, sickle-shaped, sesamoid which greatly increases the breadth of the hand and increases its efficiency in shovelling earth.

The femur of the Insectivora has a third trochanter, which is particularly well developed in *Erinaceus* and *Centetes*. The fibula is a slender bone and is usually separated from the tibia, but it is, however, fused at the distal end in *Erinaceus*. The foot is nearly always complete with five digits. In the jumping forms the metatarsals are greatly elongated.

Chiroptera

In the Chiroptera the fore limb is adapted to serve as a wing. The bones of the hand serve as the supports of a membranous patagium, which can be folded or extended like a wing.

The humerus is long and slender and there is no entepicondylar foramen. The ulna is a thin splint, the proximal end of which is free, but the distal end is fused with the radius. The bones of the proximal row of the carpus are fused, while those of the distal row are free. There are usually five digits. The thumb is free and is not enclosed in the patagium. At the end of the thumb is a strong claw. The remaining digits are elongated and are united by the patagium which they support. Both metacarpals and phalanges are elongated. In the Megachiroptera the second digit has three phalanges and ends in a small claw. In the Microchiroptera the second digit has a greatly elongated metacarpal but only a single rudimentary phalanx (completely absent in the Rhinolophidae). All the remaining digits, including the pollex, have two phalanges, except in the Phyllostomidae where there are three phalanges to the third digit. The third digit is the longest in all forms.

The hind limb in the Chiroptera is connected to the patagium. It is remarkable for the fact that it has been rotated so that the knee faces outwards, making the hallux external in position. The toes grip in a forwardly curved position. The limb is relatively small and the bones are slender. The fibula may be vestigial or absent altogether in some forms. The cartilaginous calcar projects inwards from the calcaneum and serves to support the edges of the patagium between the hind limbs.

Primates

The limbs of the Primates show no particular specialization and they conform closely to the primitive pentadactyl type. At the same time the relatively slight anatomical modification resulting in an opposable thumb is a feature of great importance.

The humerus is long and straight with a round globular head and no well marked ridges. An entepicondylar foramen is present in the Lemuroidea but is absent in the Anthropoidea. The radius and the ulna are nearly the same size and are always free, the distal head of the radius being capable of rotation round the ulna. This power of rotating the ulna is better developed in the Anthropoidea than in the Lemuroidea. The arrangement of the carpal bones follows the usual plan. There is a centrale except in the higher types (man, gorilla, and chimpanzee). There are five digits, of which the first is opposable to the others both in the hand and in the foot. Man is the only member of the order in which the hallux is not opposable.

The distal phalanges, both of the hand and the foot, have nails as a rule. Exceptions to this rule are *Daubentonia*, which has a nail only on the first digit of the hand and foot, the remainder being furnished with claws, *Tarsius*, which has claws on the second and third digits of the foot, and all other lemurs, which have the second digit of the foot furnished with a claw.

Reduction of the digits is rare, but occurs in some lemurs, e.g. the potto, where the second digit of the hand is greatly reduced.

The femur is a long and slender bone. The great trochanter is large and there is also a third trochanter in the Lemuroidea but not in the Anthropoidea. The tibia and the fibula are always separate bones, except in *Tarsius*, where they are fused at the distal ends. The arrangement of the tarsal bones follows the normal plan, except in *Tarsius* and the galagos where the navicular and the calcaneum are elongated. There are five digits, of which the first is opposable as in the hand, the only exception being man.

The group as a whole is arboreal and the lower members, when walking on the ground, move in a quadrupedal fashion, the fore and the hind limbs being of more or less equal length. In the higher forms among the Anthropoidea the adoption of a semi-upright gait is made possible by the elongation of the arms, which touch the ground and serve to maintain the balance. In man alone is the true erect position attained and

the body balanced erect on the hind limbs without any assistance or support from the arms. In man the arms are short and
the legs are long, while the foot, now only used for walking
on the ground and not required to grasp the branches of trees
in climbing, shows no opposability of the hallux.

Carnivora   In the Carnivora Fissipedia the humerus has large tuberosities, a very prominent deltoid ridge, and a deep olecranon
fossa. An entepicondylar foramen is usually present, but not
in the Canidae, Hyaenidae, and certain bears. In the carpus
in living forms the scaphoid and lunar are always fused and
there is no centrale. In the extinct Miacidae, however, which
are the most probable ancestors of the Carnivora, the scaphoid
and the lunar are separate and there is a separate centrale
present (cf. Fig. 55). In the hind limb the femur has no third
trochanter. The fibula is slender but distinct, the amount of
fusion with the tibia being very slight or none at all. The
structure of the foot is normal.

In the Carnivora Pinnipedia the structure of the limbs
shows marked adaptation to an aquatic life. Throughout the
group the bones of the arm and leg are shortened, but the
phalanges of both limbs are elongated and covered by a fold
of skin, thus converting them into paddles for the purpose
of swimming. Both the humerus and the femur are short
bones and incorporated into the musculature of the body
wall, so that in the living animal the projecting limb is only
the distal portion from the elbow. The femur suffers the
most reduction. In both the Otaridae and the Odobaenidae
it is shortened to about half the length of the tibia, while in
the Phocidae it is less than a third. In all forms, however,
the great trochanter is very prominent. The fibula is almost
equal in size to the tibia in the Phocidae, but in the other
groups it is much more slender. In the Otaridae and the
Odobaenidae the feet are bent forward when walking on land
and the animal uses them to shuffle along. In the Phocidae the
hind limbs are permanently reflexed backwards and serve only
for swimming. In all Pinnipedia the external digits (the first
and fifth) of the pes are longer than the middle ones (second,
third, and fourth).

Artiodactyla   The Artiodactyl and the Perissodactyl limbs form one of the
standard examples of the evolution of a limb, originally
unspecialized and available for many different purposes, into
one which is specialized and adapted for one particular
purpose. The unspecialized type of pentadactyl limb enables
its possessor to cope with a great variety of different circumstances and situations. But the omnibus tool can never be

quite so efficient for any particular purpose as a tool designed expressly for one purpose and for no other. Hence it is that when animals are living in circumstances where certain particular movements of limb predominate, the generalized character of the limb is lost and it evolves by natural selection into a highly modified structure adapted for one habit. The Artiodactyla and the Perissodactyla represent two large orders of mammals of widely divergent ancestry, whose limbs are specially adapted for rapid progression over flat and fairly even surfaces. They are, in fact, animals adapted to living on the open plains which constitute a very large part of the earth's surface.

The generalized type of the pentadactyl limb has been adapted in different ways in the two orders, but the end products are very similar, viz. a long slender limb, raised on the tips of the digits, which are reduced to one or two, and covered at the ends by hoofs which resist the wear of rapid travel. At the same time the range of movement of the limb is restricted to one plane parallel to the long axis of the body. The first stage in evolution was undoubtedly the adoption of the digitigrade habit, which in itself confers extra length on the limb, and at the same time gives resilience by raising the ankle joint off the ground. As the middle digits of the pentadactyl limb are usually longer than the outer ones, the adoption of the digitigrade habit will, if carried to extremes, raise the outer toes off the ground and only leave the middle ones in contact. These outer toes become useless and tend to disappear, the reduction of the outer toes moving *pari passu* with the evolution of the inner. No living Artiodactyl has five digits, but in the extinct Agriochoeridae there were five digits on the fore limbs, which were thus little different from the generalized type. A later stage can be represented by the condition seen in the Suidae at the present day, while the Cervidae and the Bovidae represent the highest development reached in the Artiodactyla.

The humerus in the Artiodactyla is large and the great tuberosity is very large. There is a well marked deltoid ridge, but never an entepicondylar foramen. In the Suidae the radius and the ulna are of nearly equal size; they fit closely together so that no rotation is possible, but they are not fused. The metacarpals of the third and fourth digits have elongated while those of the second and fifth have not, so that the toes of these two latter digits are raised well off the ground. As the evolution of the limb proceeds, the radius becomes the chief bone of the forearm; the olecranon process of the ulna always remains large, but the distal end of the bone becomes smaller and tends to fuse with the radius. In the hippopotamus the radius and the ulna are equally developed, but the distal ends have fused. The Cervidae show considerable reduction of the distal end of the ulna, but there is no extensive fusion. In the Bovidae a fusion of the two bones occurs. In the Giraffidae the distal end of the ulna is reduced to a mere splint and is fused with the radius. In the hand the articulating surfaces of the bones of the carpus at right angles to the axis of the limb become flattened, but the interlocking is retained. The elongated metacarpals of the third and fourth digits fuse together to form a 'cannon bone' in all living forms except the Suidae and Hippopotamidae. In the Camelidae the distal ends of the metacarpals have not fused and are separated from each other, so that the cannon bone is Y-shaped. The Camelidae further differ from all other Artiodactyla in having the distal ends of the metacarpals smooth instead of ridged, and also in not having the distal phalanges completely enclosed in horny hoofs. The camel is to a certain extent plantigrade, as is also the hippopotamus.

The evolution of the hind limb is in advance of the fore. There is no trace of five digits even in the known fossil forms. The femur is large, and the large size of the great trochanter corresponds with the development of the great tuberosity in the humerus. There is no third trochanter. The fibula is gradually reduced in the same way as the ulna, but the reduction has proceeded further and, except in the Suidae and Hippopotamidae, where it is slender but well developed, the remains of this bone are difficult to trace. In the foot, as in the hand, the third and fourth digits are developed and enlarged, while the remainder are reduced and tend to disappear. Fusion of the third and fourth metatarsals occurs in the same way as in the metacarpals, they are free in the Suidae and the Hippopotamidae but fused in all other living Artiodactyla. The articulating surfaces of the tarsal bones are flattened in the same way as the carpals, but the interlocking is retained. The astragalus is, however, not flat, there being a deeply grooved upper surface, while the lower surface has two facets, which articulate with the cuboid and the navicular. The navicular and the cuboid are separate in the Suidae and Hippopotamidae, but are fused in other forms. The cuboid has a notch on the upper surface, over which a process of the calcaneum slides.

Perissodactyla

The Perissodactyl limb differs from the Artiodactyl limb in many important ways, which indicate the different origin of the two orders. In the first place the axis of the Perissodactyl limb passes through the third digit and not between the third and fourth as in the Artiodactyl. Throughout the order a tendency towards functional monodactylism is evident, and a complete series, beginning with a semi-digitigrade five-toed form, can be traced culminating in the modern horse.

The humerus in the Perissodactyla is well developed. The greater tuberosity is large and the deltoid ridge is well marked. The extinct Condylarthra have an entepicondylar foramen, but this is not present in other forms. The radius is always large but the ulna varies. In those forms which have more than one toe, e.g. the tapir and the rhinoceros, the ulna is as large as the radius and is free, although it fits closely against the radius and no movement is possible between the two. In the monodactyl types, e.g. the horse, the distal end of the ulna is reduced to a mere splint and is fused with the radius. In all forms the olecranon process is large.

A series showing the evolution of the monodactyl type of limb occurring in the horse can be traced from the pentadactyl type of the Condylarthra. Beginning with the Condylarthra, *Tetraclaenodon* had five complete digits and the foot is not far removed from being plantigrade. In *Eohippus* there are four functional toes, the pollex being lost. Between *Eohippus* and the modern horse there is a large series of genera and species showing a gradual change from the fully functional tridactyl condition to that of the functional monodactylism of the modern horse. In the modern horse traces of the second and fourth digits remain as splint bones partially fused on either side of the cannon bone.

The tapir has four functional toes and the rhinoceros has three in the fore foot, but both have the main axis through the third toe. The metacarpals are not elongated and the animals are still slightly plantigrade. The bones of the carpus do not have their articulating surfaces at right angles to the axis of the limb flattened, as in the horses, and the os magnum articulates only with the scaphoid as it does in the Condylarthra, whereas in the horses the magnum articulates both with the lunar and the scaphoid.

In the hind limb of the Perissodactyla the great trochanter is very large and the third trochanter is well developed. The fibula is large and free in the tapir and rhinoceros, but it is reduced to a splint representing the proximal portion only in the horses. As in the Artiodactyla the hind foot is in a more advanced stage of evolution than the fore. The primitive five-toed condition was retained in the Condylarthra, and a similar series to that already traced in the fore limb can be traced in the hind, starting with *Tetraclaenodon* and leading up to the horse. In the tarsus of the Condylarthra the three cuneiform bones (ecto-, meso-, ento-) are all present. The ectocuneiform drops out in *Eohippus*, while in the horse there is only one cuneiform. The astragalus is deeply grooved on the upper surface as in the Artiodactyla, but the lower surface differs in being flat and articulating with the navicular alone.

In the tapir and the rhinoceros there are three toes on the hind foot. The tarsus retains two cuneiform bones, the mesocuneiform and the entocuneiform.

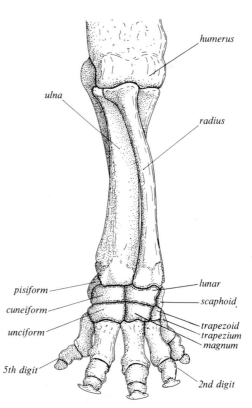

FIG. 59. Anterior view of the right fore limb of an elephant showing the permanently crossed radius and ulna, the large head of the ulna, and the arrangement of the carpal bones.

Proboscidea
(Fig. 59)

The limbs of tle elephant are straight and pillar-like in order to support the great weight of the body. In walking the limbs are not bent at the elbow or the knee but are swung from the shoulder and hip joints. Elephants walk on the tips of the toes, which are not separated externally but are marked by nails. Behind the toes both the palm of the hand and the sole of the foot have developed into large supporting cushions, so that the animal can best be described as being semi-plantigrade.

The humerus is of the same length as the radius. It is very strong and is noticeable for the great development of the ectocondylar ridge. The ulna is much larger than the radius, which is a relatively slender bone permanently crossing over the ulna. The distal end of the ulna is remarkable for being much larger than the distal end of the radius, a condition unique among mammals, but approached in the Hyracoidea. In the fore foot five toes are present, of which the middle three are larger than the outer ones. The metacarpals are not elongated. The carpals are remarkable for their flat articulating surfaces and for the fact that they do not interlock (see Fig. 59).

In the hind limb the femur is much longer than the tibia. The femur has no third trochanter. The fibula is a separate bone and articulates with the calcaneum. The number of the toes and their arrangement is similar to that in the fore foot, and the metatarsals are elongated. The tarsals, like carpals, are cuboidal in shape and do not interlock. The astragalus articulates with the navicular only.

Hyracoidea

The limbs of the Hyracoidea show certain primitive features but at the same time they resemble those of the Proboscidea very strongly in the main characters. The carpus has a centrale and the femur has a third trochanter, both of which characters may be regarded as primitive. The humerus is slightly longer than the radius. The ulna is larger than the radius and its distal end is as large as that of the radius. The radius also crosses over the ulna, as in the Proboscidea, but not to the same extent. The bones of the carpus and tarsus are cuboidal with flat articulating surfaces as in the elephant.

Sirenia

The humerus is large and has a well developed greater tuberosity. The radius and the ulna are of nearly equal size, they are free in *Halicore* but are fused at their ends in *Manatus*. At the distal ends of the radius and the ulna are large epiphyses. In *Halicore* the distal row of carpal bones have fused. There are five digits. The metacarpals and phalanges are flattened but are not elongated and there is no multiplication of phalanges as in the Cetacea.

The fore limb in the Sirenia projects farther from the body than in the Cetacea or even in the seals; the elbow joint is exposed and only the upper part of the humerus is covered by the muscles of the body wall.

There are no traces of hind limbs in the living members of this group.

Cetacea

In the Cetacea only the fore limbs are present, all traces of the hind pair having been lost. The modification of the fore limb for the purpose of assisting in swimming is here almost complete and a fin-like structure functioning in the same way as the pectoral fin of the fish has been evolved from the normal pentadactyl type. All the arm bones are short. The humerus has a very large head. The radius and the ulna are flattened and are placed side by side. The radius is usually larger than the ulna. The ulna has a well marked olecranon process. The hand is greatly modified by the fact that the normal number of phalanges may be increased to twelve or thirteen. The number of the digits is usually five, but the relative sizes vary. There is a strong tendency to the elongation of the second and third digits, which reaches its height in *Globiceps*. The carpal bones, which do not interlock, are usually arranged in two rows of three each, but in *Hyperoödon* there are five bones in the distal row. The whole of the hand is enclosed in a fold of skin and none of the fingers are free. It is doubtful whether any movement can take place between the various bones of the arm and hand, apart from a slight flexion of the paddle to modify its shape. General movement of the paddle as a whole is effected at the shoulder joint.

Xenarthra

In the Xenarthra the shape of the humerus varies with the habitat. In the sloths it is long and slender, but in the Myrmecophagidae and the Dasypodidae the ridges for the attachment of muscles are very prominent. An entepicondylar foramen is always present. The radius and the ulna are separate and in the sloths are capable of a certain amount of rotation. In the carpus there is no centrale and the scaphoid and the lunar are distinct. In the sloths the fifth digit is absent; the first digit is represented only by a small metacarpal; and there are either two clawed digits (representing the second and third) or three (representing the second, third, and fourth).

In the Myrmecophagidae all the five digits are well developed with large claws which are used for digging into termite nests; the third digit is longer than the others. Anteaters walk with the digits of the fore feet turned back so that the weight rests on the dorsal surfaces of the second, third, and fourth digits and on a thick callous pad on the extremity of the fifth digit. In the Dasypodidae there is the full number of digits but their relative sizes vary in the different forms. In *Glyptodon* and *Megatherium* the pollex is absent.

In the hind limb of the Xenarthra there is a third trochanter present only in the Dasypodidae. The tibia and fibula are of nearly equal size and, in the Dasypodidae, they are firmly ankylosed together at either end. The foot in the sloths is modified in much the same way as the hand, the second, third, and fourth digits being greatly developed, while the first and fourth are very small and vestigial, the fifth being absent altogether. In other Xenarthra the foot conforms to the usual type, it presents no special modifications and the entire plantar surface is placed on the ground when walking.

Pholidota

In the Pholidota the skeleton of the limbs shows no great modification from the normal type. The humerus in *Manis* has an entepicondylar foramen. The scaphoid and lunar are fused together. All the digits both of the fore and hind limbs, are terminated by deeply cleft, bifid, terminal phalanges. The femur has no third trochanter.

Tubulidentata

In the Tubulidentata the skeleton of the limbs does not depart widely from the normal type. The femur has a third trochanter and the humerus an entepicondylar foramen. The first digit is absent in both limbs, while all the remaining digits have pointed, clawed, ungual phalanges.

Rodentia

In the Rodentia the appearance of the humerus varies greatly. In the Muridae the deltoid ridge is very prominent, while in the beavers, the coypu, and others the ridge forms a trochanter-like process. The radius and the ulna are distinct, but are not capable of rotation. The hand nearly always has five digits but the pollex is reduced in the squirrel and is absent in the guinea pig. A centrale bone is present in some forms, but the scaphoid and the lunar are always fused. The hand of *Pedetes caffer* is interesting for the fact that there are two radial sesamoid bones jointed together to give the appearance of an extra digit. This resemblance to an extra digit is further enhanced by the distal joint bearing a nail-like structure at its extremity.

In the hind limb of the Rodentia the femur has a third trochanter, the development of which varies in different forms. In the Muridae it is one-third of the length of the femur from the proximal end while in the guinea pig (*Cavia*) and the capybara it is hardly visible at all. The fibula is separate from the tibia and is of moderate size in the beaver, the porcupine and the coypu, but in the Muridae, Pedetidae, and Dipodidae (jerboas) it is united with the tibia at its distal end and is often a very slender bone.

The foot in most Rodentia tends to be much longer than the hand. A reduction of the digits occurs in some forms. In the Pedetidae the hallux is missing, while in the capybara's foot there are only three toes, the fifth digit also being absent.

The only wide divergence from the normal type of foot, and that a most interesting one, is the condition found in the jerboas (*Dipus*). Here the three inner metatarsals are greatly elongated and have fused together, the outer metarsals are free and have also elongated but not to the same extent as the inner, in consequence of this and of the adoption of a markedly digitigrade habit the outer toes no longer touch the ground. This elongation of the metatarsal provides an extra section for the hind leg, and it is a device commonly adopted by animals requiring to move rapidly either by leaping, as does the jerboa, or by running, but the jerboa is unique among mammals in having a joint between the fused elongated metatarsals and the tarsus. This tarsometatarsal joint and the fused metatarsals are reminiscent of the condition found in the Birds.

Lagomorpha

In the Lagomorpha the humerus is long and slender and the deltoid ridge is not very prominent. The radius and ulna are distinct but are not capable of rotation. The hand has five digits but the thumb is somewhat reduced.

In the hind limb, the femur has a third trochanter which is situated close to its proximal end. The fibula is reduced to a slender slip of bone. The hallux is missing and the first metatarsal is fused with the second. All the ungual phalanges bear claws.

Index